ESCOs AROUND THE WORLD
Lessons Learned in 49 Countries

ESCOs AROUND THE WORLD
Lessons Learned in 49 Countries

Shirley J. Hansen, Ph.D.

with
Pierre Langlois
and
Paolo Bertoldi

Published 2020 by River Publishers
River Publishers
Alsbjergvej 10, 9260 Gistrup, Denmark
www.riverpublishers.com

Distributed exclusively by Routledge
4 Park Square, Milton Park, Abingdon, Oxon OX14 4RN
605 Third Avenue, New York, NY 10017, USA

Library of Congress Cataloging-in-Publication Data

Hansen, Shirley J., 1928-
　ESCOs around the world : lessons learned in 49 countries / Shirley J. Hansen with Pierre Langlois and Paolo Bertoldi. -- 1st ed.
　　p. cm.
　Includes bibliographical references and index.
　ISBN-10: 0-88173-611-2 (alk. paper)
　ISBN-13: 978-8-7702-2280-8 (electronic)
　ISBN-13: 978-1-4398-1101-6 (taylor & francis distribution : alk. paper)
　1. Energy industries--Case studies. I. Langlois, Pierre. II. Bertoldi, Paolo. III. Title.
　HD9502.A2.H34 2009
　333.79--dc22

2009000223

ESCOs around the world : lessons learned in 49 countries / Shirley J. Hansen with Pierre Langlois and Paolo Bertoldi.
First published by Fairmont Press in 2009.

©2009 River Publishers. All rights reserved. No part of this publication may be reproduced, stored in a retrieval systems, or transmitted in any form or by any means, mechanical, photocopying, recording or otherwise, without prior written permission of the publishers.

Routledge is an imprint of the Taylor & Francis Group, an informa business

0-88173-611-2 (The Fairmont Press, Inc.)
978-1-4398-1101-6 (print)
978-8-7702-2280-8 (online)
978-1-0031-5149-4 (ebook master)

While every effort is made to provide dependable information, the publisher, authors, and editors cannot be held responsible for any errors or omissions.

Table of Contents

Foreword vii

Chapter 1.   ESCO Development 1

Chapter 2.   **Western Europe**
Germany—Italy—Greece—United Kingdom—Ireland—France—Austria—Belgium—The Netherlands—Luxembourg—Finland—Sweden—Denmark—Portugal—Spain 15

Chapter 3.   **Eastern Europe**
Czech Republic—Poland—Romania—Hungary—Slovakia—Croatia—Russia—Bulgaria 69

Chapter 4.   **Africa**
Tunisia—Algeria—Morocco—Cote d'Ivoire—Kenya—South Africa 101

Chapter 5.   **The Middle East**
Lebanon—Israel — Turkey—Egypt 125

Chapter 6.   **Asia**
Japan—Thailand—Malaysia—Vietnam—India—Philippines—Republic of Korea—China 147

Chapter 7.   **North America**
Canada—United States—Mexico 229

Chapter 8.   **South America**
Chile—Uruguay—Brazil 267

Chapter 9.   **Down Under**
New Zealand—Australia 305

Chapter 10. **The Global Picture** 313

References 341

Appendix A: Contributing Authors 353

Appendix B: International Partnership for Energy Efficiency Cooperation Declaration 357

Index 363

Foreword

Imagine an industry where 25-50 percent return on investment is common. An industry, which can cost-effectively reduce pollution and bring us closer to our sustainability goals. An industry where customers are offered reduced operating costs and new equipment without front-end capital expenses. An industry where the project costs are paid for out of avoided utility costs—guaranteed. An industry where targeted projects can reduce carbon emissions; thus, avoiding important impacts on climate change. The appeal is enormous; so it is not surprising that the energy performance contracting industry and the energy service companies (ESCOs) which offer all this, have been growing rapidly.

In fact, the appeal is so great that one wonders why there is not a crush of new firms striving to be ESCOs on every street corner in the world, and where potential beneficiaries are just running around to find such a service provider. Unfortunately, there are also barriers and difficulties that work against such an industry.

Given this background, it seems inevitable that questions regarding the status of the industry are frequently asked. Such questions as: How widespread is the worldwide ESCO industry today? Is it true there are fewer ESCOs in the US today than ten years ago? If so, why? What about the growth internationally?

What are the different barriers, difficulties and limitations to this industry in various countries? And where are the exciting new opportunities around the world? Do lower rate schedules due to subsidized tariffs still limit ESCO activity? To what extent are potential clients, which have poor financial ratings, a factor? Are uncertainties about ownership, potential effects of privatization, or the lack of a paying customer base holding back ESCO industry growth in developing and emerging countries? Has the liberalization of the business climate overcome the inefficiencies and ineffectiveness of publicly controlled economies or sectors?

This book offers an assessment of today's ESCO industry around the world. Current ESCO trends are assessed by geographical regions and specific countries. Analyses of opportunities and barriers/problems are presented by in-country people who know their ESCO industry.

We are indebted to exceptional in-country experts, who have provided us critical insight into local conditions. These contributing authors, with some reference to their credentials, are cited in the respective chapters. Short biographies are provided in an Appendix A.

Drawing on his work around the world as president of Econoler and Econoler International, Pierre Langlois has played a vital role in preparing this book. He has put us in contact with key individuals in specific countries, served as a contributing author regarding Canada and reviewed key portions of the manuscript. As a principal author of the European Commission's ESCO report of 2007, Paolo Bertoldi, DG, Joint Research Center of the European Commission, provided the initial idea for this book and provided the basis for the European material in Chapters 2 and 3.

Hours and hours of work went into editing the material from the respective countries to provide greater consistency to the presentations and to smooth out some English translations. We are incredibly fortunate that we had someone with great technical writing capabilities to serve this need. Drawing on his engineering degree and years in radio, TV and public relations, we are deeply indebted to Jim Hansen, President of Hansen Associates, for the flow from report to report and its readability.

The caliber of the contributing authors and reviewers assure the reader a reliable assessment of ESCO conditions in various parts of the world. While some omissions and discrepancies are inevitable and perspectives will vary, the content of this book represents the best thinking of the current status of the worldwide ESCO industry.

Chapter 1
ESCO Development

Those of us, who have advocated energy performance contracting (EPC) for decades, can take great joy in reading the reports in this book detailing ESCO business development from around the world. Many of us have been well aware of energy performance contracting's economic and environmental benefits and now receive incredible gratification in seeing it recognized by increasing numbers of people across the globe.

The difficulties inherent to EPC, which are depicted by an impressive number of recognized figures in their respective countries who have contributed to this effort, could have discouraged many embryonic efforts. The absence of support, favorable legal frameworks, or limited financing constituted major barriers and could have defeated young and emerging ESCO industries. Sometimes governments were there to help. Too often, they were major hindrances—either unwittingly or deliberately.

But ESCOs offered a good thing—eliminating many of the traditional barriers for energy efficiency (EE) projects through reduced operating costs without capital expenditure to industry, and skills that freed up critical money for schools and hospitals. Add to that an industry that preserves precious human and natural resources and helps clean up our environment while making money. The story is a good one and, despite its problems, the ESCO industry has prevailed.

We have reached into the far corners of the world to get our trusted colleagues to tell the story of their industry's development in their respective countries—and in their own words. Sometimes, we had to edit the material to offer more consistency among the different reports and chapters. But we have tried to keep the flavor, and undeniable charm, for all to enjoy. In so doing, we kept the uniqueness of the telling, which kept us from forgetting that these are first-hand accounts of failures and success, struggles and triumphs.

To set the framework for this telling adequately, we need to take a few pages to describe the energy climate into which the EPC concept was born and how it has evolved into the industry it is today worldwide.

THE "ENERGY" AWAKENING

Nearly four decades ago, we awoke to find that "energy" was not necessarily tied to words like *static* or *kinetic*. Amid leaky buildings and wasteful practices, we discovered that energy was, and is, the lifeblood of our economies and vital to our ways of life. "Blood" that has become even more critical with the advent of computers and other technical innovations.

Through the years, increased populations and their demands for improved life styles have put even greater pressure on our current energy supplies and the ways we use energy. A trend that is expected to exacerbate the problem if left unattended.

Growing Energy Needs

Between 2004 and 2030, the global GDP is expected to more than double worldwide. To serve this growth, the International Energy Agency (IEA) projects a 53 percent growth in energy demand during this period, which will require 5.9 billion of tons of oil equivalent (btoe).

The developing countries (non-OECD) are expected to contribute 80 percent of the world's economic growth from 2004 to 2030. Expectations that this growth and a modest increase in economic prosperity will nearly double this sector's energy demand, adding 4.2 btoe. China and India are expected to lead this demand as they become the two greatest energy consuming nations in the world.

In many parts of the world, the energy community, especially utilities, is experiencing less state control and is slowly moving towards free markets. The strain caused by this transformation is enormous and affects all areas: finance, energy services, performance, administrative regulations, taxes, marketing and information technology. Within a short time, companies, which are preparing themselves for worldwide markets, have found that EPC is a global game.

The key growth drivers for the global energy market have been the rapidly increasing energy consumption in the developing world and growing concern about our environment. Energy demand in the developing countries of Asia, which includes India and China, has been projected to increase as much as 100 percent by 2023.

Environmental Implications

Amid concerns regarding energy security and rising costs, we have come to recognize the negative impact our current energy usage has on the environment, especially the consequences of burning fossil fuels have on global warming. It is accepted that the current levels of atmospheric greenhouse gas (GHG) can be primarily attributed to the greater energy use by the industrialized nations, generally characterized as the Organization of Economic and Cooperative Development (OECD) countries.

The projected growth in non-OCED countries' (developing and transitional economies) energy use raises the specter of tremendous environmental consequences, both damaging and costly. IEA now projects that energy use in developing countries (non-OECD) will soon surpass the CO_2 and GHG contributions from OECD countries.

At the same time, foreseeable energy demands continue to open up incredible opportunities for firms offering EE equipment, support and services.

The Energy Efficiency Answer

Improving the efficiency of energy use is recognized as the most cost-effective way, by far, to gain increased energy security, improve industrial profitability, assure greater competitiveness, and reduce the overall impact on climate change.

Learning the most effective ways to reduce energy consumption has not come easily. Since the energy price shocks of the 1970s, we have grappled with ways to cut consumption. Through the years, we have gradually learned what measures most cost-effectively save energy. In the process, we have also looked for ways to get this information and the financing it requires, to the end users.

As we realized that EE actually is one of the most interesting financial opportunities in the market today, we also recognized that EE is really an investment; not an expense.

EE costs vary by technology and among countries. These costs are frequently one-quarter to one-half the comparable costs of acquiring additional energy supply. In addition, increased energy supply also increases global pollution. EE can actually make money while reducing GHG.

While the benefits of EE are clear, implementing them on a large scale has been difficult. The energy cost saving measures are technically and logistically diverse and often small in scope. They do not compete well for capital against capacity or market expansion. From the financiers' perspective, high transaction costs and perceived risks make EE investments less attractive. If we are to realize these opportunities, it would be very desirable to have an entity that can aggregate projects, demonstrate technical expertise, manage/mitigate associated risks and guarantee results—an energy service company, an ESCO.

The IEA's 2006 study, *Light's Labours Lost*, provides an excellent example of EE benefits. It found that should energy users install only efficient lamps, ballasts and controls, significant money could be saved over the life cycle of the lighting service. Such an investment would cause the global demand for electricity to drop substantially. In fact, the savings would give our increasing demand a free ride. The lighting model would result in total demand remaining unchanged from 2005 to 2030 (IES 2006a). The bottom line: avoided cost in total lighting expenditures would be USD 2.6 trillion and the avoided emissions would be 16,000 million tons of CO_2 emissions.

Such findings led to the G8 countries, China, India, South Korea and the European community to establish the International Partnership for Energy Efficiency Cooperation at the Energy ministerial meeting on 8 June 2008 in Heiligendamm. At this summit, the newly formed IPEEC declared in part:

- The Partnership offers to the G8 and other interested countries a flexible forum for high level policy discussion, regular strategic cooperation and exchanges focused exclusively on energy efficiency. It will support the ongoing work of the participating countries and relevant organization to promote energy efficiency. It will be a supplementary and complementary instrument to the United Nations Framework Convention on Climate Change process.

  The objectives of the Partnership:
  — Secure a clearer picture of international action on energy efficiency;
  — Enable the development of a shared and strategic view covering these activities; and
  — Identify jointly the possible collaboration actions and maximizing the impact and synergies of their individual national actions.

The following day (9 June 2008), the ICEEP made a declaration, which emphasized the need for global cooperation in the field of EE. This declaration stressed 14 points which underscore the value of EE on a global scale. When financing is repeatedly cited as the most frequent reason for not doing energy efficiency work, the ICEEP declaration (presented in its entirety in Appendix B) offers an excellent platform for an even stronger movement in ESCO development.

THE EMERGING ESCO INDUSTRY

Once it was demonstrated that EE makes money, ideas to capture this concept as a business proposition began to emerge. The first effort came from Scallop Thermal, a division of Royal Dutch Shell. Scallop took an idea, which had been used on the supply side of the meter for nearly 100 years, and gave it meaning on the demand side. Compagnie Générale de Chauffe (CGC) had for years been guaranteeing savings from its work in district heating. Now Scallop took this concept in the late 1970s to the UK and US offering to deliver conditioned space to its customers for 90 percent of their current utility bills. Scallop had determined a way to effectively manage facility conditions for less than 90 percent of the baseline and the concept of "shared savings" was born.

Ironically, in both the UK and US the governments opposed the concept, which was later to be known as performance contracting. In the UK, a little known accounting officer named John Majors, later to serve as Prime Minister, created government language to permit such a program. In the US, an embryonic energy services industry got laws passed state by state to make it happen.

The struggles, however, to establish this young industry was a long way from over. The CGC concept, as developed by Scallop, provided that each party would share a predetermined percentage split of the energy financial savings. During the life of the contract, the ESCO expected its percentage of the cost savings to cover all the costs it had incurred, plus deliver a profit. This concept worked quite well as long as the energy prices stayed the same or escalated.

But in the mid 80s, oil prices dropped and it took longer than expected for an ESCO to recover its costs. With markedly lower energy prices, paybacks became longer than some contracts. Firms could not meet their payments to suppliers or financial backers. ESCOs closed their doors; and in the process, defaulted on their commitments to their shared savings partners. "Shared savings" was in trouble -- and the process became tainted by lawsuits and suppliers' efforts to recoup some of their expenses. At the same time, facilities managers valiantly tried to explain losses previously guaranteed.

Fortunately, many ESCOs persisted in their efforts to make the new concept work. Some agreements continued to show savings benefits to both parties. Of even greater importance, several companies, which had guaranteed the savings, made good on those guarantees.

In spite of this tenuous start, the "shared savings" industry survived, but its character changed dramatically. Those, who were supplying the financial backing and/or equipment, recognized the risk of basing contracts on future energy prices. With uncertainty in the industry and greater uncertainty in energy pricing, risk levels grew and interest rates went up accordingly. The use of the original "shared savings" agreements shrank to approximately 5 percent of the US ESCO market.

In its place, new names, new terms, new types of agreements, and a very different financing mechanisms emerged. Perhaps to respond to the negativity that had been generated around "shared savings," the industry focus turned to guaranteed energy *performance*. The term, *energy performance contracting (EPC)*, became the favored name as a new model, "guaranteed savings" was established.

In Europe, the EPC also became the popular term for the concept, but the model remained heavily focused on shared savings.

In the midst of the shared savings disaster, the idea of guaranteeing the amount of energy to be saved took hold. Secondarily, the ESCOs guaranteed that the value of that energy would be sufficient to meet the customer's debt service requirements provided the price of energy did not go below a floor price as specified in the contract.

From its shaky beginnings to its near death when oil prices plunged in 1986, a strong performance contracting industry emerged in the North America and provided the impetus for strong growth around the world.

Part of the appeal of EPC was, and is, in the models' ability to package services, equipment and related measures to create a project size more attractive to financiers. Another attractive aspect of EPC is the mechanism it provides for the ESCO to use *future* years of predicted energy cost savings *now*. For example, a project predicted to save $1 million per year on a ten-year contract can make $10 million immediately available for a project. $10 million can buy a lot of services and equipment—and savings.

As EPC established itself as an accepted way to provide or enable financing for EE projects, it reinforced the efforts in Europe and inspired development in Asia. The ESCO industry began moving aggressively in all corners of the world. Soon there were conferences designed to spread the word of effective best practices and to cement the existence of this new industry. Indicative of its growing popularity, the 2nd European conference, called ESCO Europe 2005, and the 1st Asia ESCO Conference in Bangkok were held within weeks of each other in the fall of 2005. Some of this growth has been documented in the 2005 report by the European Commission. In 2007, the EC updated its report, which provided an overview of the ESCO status in European countries.

As noted in these reports, and others on ESCO development, gathering information and making comparisons of ESCO markets is limited by the fact that the notion of an *energy service company, or ESCO,* is understood differently from one country to another, and sometimes used differently by experts even in the same country. The problem with definitions has been highlighted in many forums and by numerous experts and business actors. The authors have, therefore, accepted for this book, the following terms as issued by an EC directive:

> "**energy service company**" (ESCO), a natural, or legal person that delivers energy services and/or other energy efficiency improvement measures in a user's facility or premises, and accepts some degree of financial risk in so doing. The payment for the services delivered is based (either wholly or in part) on the achievement of energy efficiency improvements and on the meeting of the other agreed upon performance criteria;

> "**energy performance contracting**" (EPC), a contractual arrangement between the beneficiary and the provider (normally an ESCO) of an energy efficiency improvement measure, where investments in that measure are paid for in relation to a contractually agreed level of energy efficiency improvement.

Put more simply, an ESCO guarantees its energy savings performance. Parallel organizations, such as energy service providers (ESPs), may offer the same service, but do not offer the guaranteed results.

In the 21st century, the maturing of an industry is increasingly marked by the formation of industry associations. An affirmation of the growing ESCO industry came in 2007 when a networking meeting of the Asian country associations was held in Beijing just prior to the 2nd Asia ESCO Conference.

Once described as "alternative financing" by the US federal government, EPC is no longer an *alternative*; it is an accepted way of doing business. The ESCO industry delivers EE expertise, financing and ways to meet environmental mandates. The EPC concept is now pervasive and persistent around the world. But it has not happened overnight and its development has been uneven. Many different barriers still limit the potential benefits that EPC could deliver around the world.

ESCO Services

EPC continues to change and evolve. Services are more sweeping and financial models are more flexible. ESCOs today offer a broad range of retail energy services, including:
- engineering feasibility studies, audits and investment grade audits
- equipment acquisition and installation
- load management
- supply; power marketing
- facilities management and water management
- risk management
- automated meter reading
- indoor air quality services
- energy information management
- training and awareness services
- sustainability support and environmental compliance
- measurement and verification of savings
- guaranteed results.

Following the patterns that evolved in the telecommunications industry, ESCOs are showing an increasing tendency to unbundled and bundle these services... offering several or all of the above to their customers. Ultimately, ESCOs are apt to be selling conditioned floor space, which will provide ESCOs and end users a more effective and efficient means of guaranteeing the return on investment. Such an advent would bring us full circle back to the first EPC developed in the US by Scallop at Hanneman Hospital: Chauffage—an integrated supply/use efficiency solution.

In the meantime, those just now considering EPC -- as a consumer, a financier, or an engineering company -- have a history rich in experience to draw upon. We now have the ability to look at projects, which have repeatedly put the theory of EPC into practice. We have learned what works.

More importantly, we have also painfully learned what doesn't! This book is designed to encapsulate these experiences and give the reader an important reference source.

Financial Models

The two dominant EPC models in the world are shared savings and guaranteed savings. With some minor modifications, the shared savings model is currently practiced in Europe, and to a lesser extent in the North America. For about 10 years, shared savings was the only type of EPC offered by North American (US and Canadian) ESCOs.

The primary characteristics of shared saving can be summarized by the following:
- Customer and the ESCO share a predetermined percentage split of the energy *cost* savings;
- ESCOs carry both the performance risk and the credit risk;
- Financing for the customer is often off balance sheet;
- Equipment, which is often leased, is "owned" by the ESCO for the duration of the contract (ownership is usually transferred to the owner at contract end);
- Increased risks, such as the uncertainty of energy prices, cause the cost of money to be higher;
- Unless special safeguards are implemented, customers have greater payment exposure if energy prices or savings increase; and
- As all the project costs are recognized as a service, they are fully deductible in many countries for the duration of the agreement.

The economic viability of shared savings rests on the price of energy. As long as energy prices stay the same or go up, the project will typically pay for itself.

As mentioned before, after the mid-1980s saw energy prices drop, which prompted the development of a model that no longer relied so completely on the price of energy to establish the project's economic viability. ESCOs in North America shifted to guaranteeing the amount of *energy* that would be saved, and further guaranteed that the value of that energy would be sufficient to meet the customer's debt service obligations so long as the price of energy did not fall below a stipulated floor price.

The significant characteristics of guaranteed savings can be summarized by the following:
- The amount of *energy* saved is guaranteed, as long as the operation

remains similar to the period preceding the project implementation;
- Value of energy saved is guaranteed to meet debt service obligations down to a stipulated floor price;
- Owners carry the credit risk;
- Risks to owners and ESCOs are less than with shared savings;
- Less of the investment package goes to financing costs; and
- Tax-exempt institutions, in countries that provide for this tax provision, can use their legal status for much lower interest rates.

While shared savings remains the dominant model in Europe, in North America over 90 percent of the EPCs are currently structured for guaranteed savings with the owner typically accepting the debt through third party financing. In preparing the material for this book, one of the drivers was to ascertain which model is most prevalent in the various countries around the world.

The typical cash flow of these two financing models is shown in the following figure. In analyzing this cash flow, there are two distinguishing characteristics that should be noted.

First, a formal contractual arrangement between the ESCO and a lender does not always exist. This relationship may vary from an informal understanding to a very detailed agreement. When informal relationships are established, certain conditions are typically understood and may include:
- Customer pre-qualification criteria;
- Project parameters;
- Stream-lined lending procedures, which have been cooperatively developed; and
- Special interest rates.

The second distinguishing characteristic appears in shared savings. In this case, the customer has no relationship with the financing institution and has little or no specific interest in seeing that the loan is repaid. Since all the savings must happen in the customer's facility and/or process, this factor further raises the risks to the ESCO and the financier.

Reasons do exist to encourage the shared savings model. One major reason is the difficulty customers in transitional economies have in satisfying the bank's criteria for creditworthiness. Another reason is the fact that a new concept, such as EPC, is easier to establish in a country if the ESCO's customer does not have to incur debt. A third reasons is the desire

ESCO Development 11

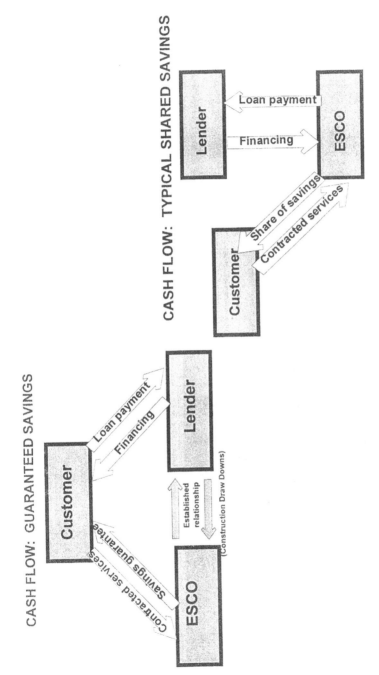

Figure 1-1. Dominant Performance Contracting Financial Models

on the part of some energy end users to avoid incurring further debt, or going through the political/legal procedures to do so.

Shared savings, however, relies heavily on ESCO borrowing capacities and this presents a serious difficulty for small and even big ESCOs which lack access to financial resources rapidly or over time. After incurring debt on even a limited number of projects, an ESCO is apt to find it is too highly leveraged to obtain financing for the implementation of more projects. This is a key factor in hampering industry growth. To satisfy the ESCO's needs and to continue to avoid an untenable debt load, some ESCOs have turned to an emerging financial model which establishes Special Purpose Entities (SPE), or in some countries it is called a Special Purpose Vehicle (SPV). The cash flow for this model is shown in Figure 1-2.

In this model, the SPV collects the revenues and pays the financier. Typically, the financial house and the ESCO are joint owners of the SPV.

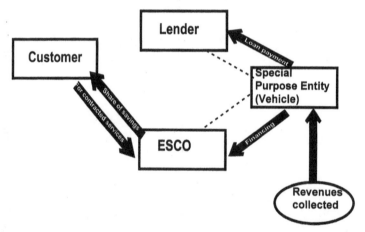

Figure 1-2. The Special Purpose Entity Model

Moving up the Value Chain

The dominant financial models are typically applied to comprehensive demand-side management (DSM) at the second level in the value chain as shown in Figure 1-3. Generally depicted at the next higher level in the value chain are efficiencies in supply, which might be district heating efficiencies, combined heat and power (cogeneration), or the implementation of stationary fuel cells. It is placed above comprehensive EE services only because the dollar amounts can be greater for work on the supply side of the meter.

Business Solution

Chauffage

Integrated Solution

Supply Efficiencies

Comprehensive Demand Efficiencies

Single Measure

Figure 1-3. The EE Value Chain

When, in addition to some demand side services, the same ESCO provides a broader range of supply acquisition services, such as cogeneration or distributed generation, the package is referred to as an integrated solution.

The terms, integrated solutions and chauffage, are sometimes used interchangeably, but chauffage generally refers to a greater value-added approach. Integrated solutions may simply refer to a supply contract and a demand contract offered by the same ESCO, while chauffage offers conditioned space at a specified price per square foot (or square meter). In such a case, the ESCO manages all supply and demand efficiencies. In practice, ESCOs sometimes focus only on supply efficiencies and refer to the contract as "chauffage." It may include some type of ownership of a part, or the totality, of HVAC systems by the ESCO. The contract typically provides for some means of making adjustments for energy prices on an annual basis.

The ultimate value-added on the supply chain is the business solutions approach. Typically that approach allows an ESCO to propose solutions that make prudent business sense, which may go beyond reduced energy consumption. The ESCO may provide services beyond energy efficiencies, wherein the energy cost savings may help defray the costs of this additional work. In other instances, the work may actually increase energy costs, but lower the energy cost per unit of product through process efficiencies.

And Then What? EPC Around the World

The projected huge escalation in energy usage noted earlier in this chapter can be very daunting. However, it can also be viewed as an incredible opportunity for EE and more particularly, ESCOs. The following chapters address these opportunities by regions of the world with specific sections devoted to certain countries. This material was gathered from in-country experts and gives the reader a critical insight as to the way in which the industry is progressing. It was impossible to cover all the countries where EPC has been introduced, or is currently emerging; however, we believe the reviews presented in this book will provide the reader a very good general understanding of the global EPC situation.

As the reader progresses through the book, it will become evident that ESCO industry development in each country has its own unique characteristics. The factors that encourage and/or discourage EPC vary in degree and in impact. There are, however, a number of common threads that seem to weave their way around the world. The final chapter of the book summarizes these common threads and offers a few thoughts to those interested in furthering this intriguing EPC concept.

Herein is the very first look at ESCO development around the world.

Chapter 2
Western Europe

*T*he first companies offering services in the energy field and applying the ESCO concept appeared in Europe as long ago as the 1800s. The cradle of these so called "operators" or "managers" was France. The concept gradually moved to other European countries, but really boomed in North America during the 20th century.

EPC in Europe

Energy performance contracting (EPC), as it is referred to in Europe, has its roots in the early work of Compagne Générale de Chauff (CGC) in France. From the guaranteed supply savings in district heating systems through greater energy efficiency offered by CGC, an energy service industry emerged. Integrated energy efficiency solutions, which considered efficiencies on both sides of the meter slowly spread through Europe in the 1980s. EPC has had an uneven pattern of development as a quick overview of Europe reveals.

Germany is considered the largest and most advanced market, with France and the UK, Spain and Italy following closely behind. At the same time, ESCO markets began to emerge in Central and Eastern Europe, too. The development of the ESCO market in Eastern Europe is addressed in Chapter 3.

The total European market potential has been estimated to be at least €5-10 billion per annum with an expected €25 billion in the long term. Investing in energy efficiency with the help of ESCOs is, in principle, a particularly profitable business in any European country; however, actual profitability depends on many factors and can be curbed by a wide array of barriers. ESCOs are profit oriented businesses and should not be expected to intervene in areas that are too risky or do not offer a profit.

The majority of projects developed by energy service companies in Europe have been undertaken in the public sector, where the model of public-private-partnership (PPP) is one of the most effective tools to boost energy efficiency. The most common technologies so far have been cogeneration, public lighting, heating and cooling, ventilation and energy management systems.

The use of cogeneration, or combined heat and power (CHP), in Europe constitutes a substantial portion of ESCO business. CHP offers the simultaneous production of heat and power, and can therefore offer energy savings and avoid CO_2 emissions compared with separate production of heat and power. In addition, development in the use of fuels used in CHP applications has shown a trend toward cleaner fuels. Nearly 40 percent of the electricity produced from cogeneration is produced for public supply purposes, often in connection with district heating networks.

The increasing natural gas prices have, however, had an adverse affect on the CHP market in some European states. Barriers to accessing national electricity grids to sell surplus electricity, and relatively high start-up costs have also depressed the market.

According to the *Energy Business Reports,* "Combined Heat and Power Developments in Europe," of August 7, 2008, government actions are helping to reduce the barriers to CHP.

The CHP directive on promotion of high-efficiency cogeneration is expected to start having an effect. It encourages member states to promote CHP uptake and helps them to overcome the current barriers hindering progress. It does not set targets, but instead requires member states to carry out analyses of their potential for high-efficiency cogeneration. A number of EU member states have introduced laws or other support mechanisms to promote new CHP. Despite these measures, there remain substantial differences in the level of CHP across the EU. Countries with a high market penetration of CHP electricity include Denmark and the Netherlands, while poor infrastructure for natural gas and less demand for heat has historically hindered CHP development in countries like Portugal and Ireland.

The *European ESCO Status Report 2005* listed 9 major EPC barriers, which still prevail and have worldwide implications:
- Low awareness, lack of information and/or trust and skepticism on the clients' side;
- Limited understanding of energy efficiency opportunities and EPC;
- Small project size and high transaction costs, which has discouraged financing and ESCO business;
- High perceived technical and business risks;
- Legal and regulatory frameworks not compatible with energy effi-

ciency investments, for instance non-supportive procurement rules;
- Limited understanding of measurement and verification protocols for assuring performance guarantees;
- Administrative hurdles, such as complicated procedures, high transaction costs, split incentives, and aversion to opt-out energy management tasks;
- Lack of motivation because energy costs are only a small fraction of total costs; and
- Limited governmental support for EPC.

In order to overcome some or all of the above hurdles, the authors of the *European ESCO Status Report* suggested certain soft actions, including:
- Increasing dissemination of information about ESCO services and projects;
- Launching an accreditation system for ESCOs (proving the quality and reliability of services);
- Developing financing sources;
- Standardized saving measurement and verification; and
- Ensuring that governments take the lead with measures in public buildings.

The above barriers and enablers are explained (including details and examples) in the *European ESCO Status Report 2005*, in which additional literature and references can be found. The more recent *ESCO Update Report of 2007* looked further at existing barriers and success factors. The European Union requires member states to submit a national energy efficiency plan every three years. Many of the plans address the reporting country's ESCO situation. The reader is referred to the European Union's NEEP web site to get the current status of energy efficiency and ESCOs in the member countries.

Much of the remainder of this chapter is based on an updated version of the 2007 report, which was authored by Paolo Bertoldi, Benigna Boza-Kiss and Silvia Rezessy. We are indebted to Anees Iqbal for the valuable information he supplied on the United Kingdom. Reaching back to the inception of the performance contracting concept. Jérôme Adnot and Frédéric Rosenstein trace the evolution in France to the country's current unique "ESCO" classification system. Current ESCO conditions in Belgium are offered by Leiven Vanstraelen, who is the manageing director of Fedesco.

WESTERN EUROPE ESCO MARKET

This chapter is essentially a look at Western Europe as a complex but open, thus permeable market. A description and analysis of several national ESCO markets in Western Europe are offered.

The country overviews in this and subsequent chapters have similar structures to help the reader navigate through the information, but they are separate documents and function as brief reports. Within a given country's discussion, an occasional reference to other countries may be made. This is done to reveal commonalities and differences, and because one country's market is often strongly related to others.

In general, firm numbers on the size of the ESCO market are not clearly evident. Where figures exist, they are too often dated and, therefore, are not suitable for the purposes of this book. The potential of the EPC markets were more often found or estimated by experts and other interviewees contacted for the book.

Furthermore, when available, the most important barriers are presented, with an indication in some instances of what needs to be done in order to successfully overcome the obstacles and enhance the ESCO markets. Finally, trends and the expected future of the ESCO industry in a given country are offered.

The order in which countries are presented does not indicate any prioritization or level of importance, but should be considered as a pure list of countries.

GERMANY

The German ESCO market is often celebrated as the most established energy service industry in the European Union. It is among the oldest ESCO markets in Europe, emerging in the early 1990s, and has experienced a constant expansion ever since. In spite of the early start, the continued increase in activity and the overall success of the German ESCOs, significant market potential still exists.

ESCO Market Size and Activity

The overall number of ESCOs and ESCO-like companies in Germany is estimated to be around 500. The majority of these companies offer energy supply contracting (particularly heat delivery services) and operations contracting. The number of companies, which are actually offering services through energy performance contracting (EPC) as defined in Chapter 1, is only a fraction of the total figure, around 50. ESCOs with more than one referenced EPC project are in the range of 20. Small and large local companies, including former municipal utilities and multinational companies, are active in the market. Furthermore, the four largest energy companies all have daughter companies carrying out various contracting activities, of which one is particularly active in the EPC business.

There are two associations helping the ESCO sector via a range of activities. The newly established ESCO Forum represents the larger ESCOs. The ESCO Forum is a recent merger of the former Bundesverband Privatwirtschaftlicher Energie-Contracting-Unternehmen e.V. (PECU) and the Contracting Forum of the German Electrical and Electronic Manufacturers' Association (Zentralverband Elektrotechnik- und Elektronikindustrie e.V.—ZVEI). In 2007, the ESCO Forum had 26 members. On the other hand, the Verband für Wärmelieferung (VfW) is an association of mostly smaller heat delivery service suppliers. As of 2005, VfW had 230 members, of which 197 had contracting projects.

The total number of running ESCO contracts is estimated at 50,000. In 2005, Flauger referred to a total potential of 1.3 million projects in Germany. In 2005, the total turnover of the members of VfW amounted to €1.04 billion (including energy revenues). New investment amounted to €510 million. 83 percent of the contracting activity was energy supply contracting, 8 percent EPC, 5 percent management of technical equipment, and 4 percent pure third-party financing. According to other sources, the share of EPC in the market is around 15-20 percent.

The EPC market in Germany had a total investment value of €750 million by 2006. The market potential is estimated to be about €2 billion in the public sector alone, which includes energy turnover, and corresponds to an annual potential of €350 million monetary savings volume from energy savings, according to Berliner Energieagentur GmbH. The most common contract model is the guaranteed savings scheme. Excess savings are shared between the client and the ESCO following a previously agreed percentage.

The average payback time of ESCO projects is 5-15 years, with the municipal sector tending toward longer projects. The municipal growth is principally due to the trust that has been developed to a large extent and greater reliance on outsourcing. Industry still appears averse to long-term contracts, thus shorter contracts dominate. Therefore, the payback times are also shorter, around three years. Average savings of EPC contracts in Germany are in the range of 10 to 38 percent for 0.2-2,000 MWh/EPC contracts respectively.

Growth Factors

In the beginning of the nineties, only a limited number of EPC projects were initiated, no standard documents were available, and doubts about the trustworthiness of ESCOs, their reliability, and the correct value of contracts hindered the sector in Germany.

The establishment of the Energy Saving Partnership (ESP) in 1995 in Berlin is considered an important step in establishing the energy efficiency market in the public sector in Germany. Under the ESP scheme, buildings are bundled into pools in order to decrease transaction costs. 21 pools had been contracted by ESCOs by 2006, encompassing over 1300 buildings altogether. A notable number of EPC projects have been realized in Hessen, North-Rhine-Westfalia, and Bavaria. There are, however, less or no activities in other regions, such as Lower Saxony and in the Eastern Länder.

In the meantime, an additional scheme, called "Energy Saving Partnership Plus" is being set up, in order to embrace building and construction measures, including heat insulation, and window replacement. This scheme is based on the existing one and expands its application by also including work on the building shell, instead of the typical focus on energy system improvements (equipment and control engineering). This comprehensive approach is expected to attract new customers from different sectors, such as industry, hospitals, offices, and housing.

According to some experts the share of demand-side EPC is actually decreasing in Germany, but the integration of demand-side energy efficiency measures with supply-side oriented contracting is gaining importance. The provision of energy supply services is successful with private sector buildings. The ESCO market is projected to be further boosted in Germany by the anticipated expansion of combined heat and power (cogeneration).

The successful ESCO industry in Germany is the consequence of a mixture of favorable conditions, but it is mainly the result of local political support and individual drivers. A large number of municipal projects, many of which are supported by the energy agencies, have a strong demonstration effect and act as multipliers among other sectors, most notably the commercial sector.

Besides the large private ESCO sector, Germany is the homeland of the so-called "Interacting Model" or Public Internal Performance Commitments (PICO). In the PICO model, one department in the administration acts as a unit similar to an ESCO in function for another. The ESCO department organizes, finances and implements energy efficiency costs savings. However, these projects lack the energy savings guarantee, because there are no sanction mechanisms within a single organization (even though PICO includes saving targets). This can result in lower effectiveness of the investments. Nevertheless, this scheme offers improvements mostly through a fund made up of municipal money, and using existing know-how. This allows a company to undertake less profitable projects, which often yield increased activity for energy savings.

Furthermore, increases in energy prices since the liberalization of the electricity market is considered to be one of the most important triggers for the German ESCO sector. As a result of liberalization, energy prices dropped significantly between 1999 and 2001, but at the same time energy taxes increased, and in the period 2002-2006 energy prices almost doubled. Some ESCOs have reported that the energy taxes are one of the most effective political measures for energy efficiency.

Other keys to the evolution of the ESCO industry were the establishment of standard procedures and documents, such as model contracts; an energy performance retrofitting model; and a standard procurement procedure. In addition, contracting guidelines by the federal states of Hessen and Berlin were developed. Today, there are approximately 7 different model contracts.

Supportive Government Actions

The German government supports investments aiming at sustainable energy use and energy conservation through various financial and technical mechanisms, including research and development programs, loan/funding schemes, and incentive programs for renewable energy. Additionally non-governmental programs also exist (such as credit programs by eco-banks, for instance kWf, or boiler replacement by utilities), which complement ESCOs' work in the residential sector. Energy agencies at national, regional and local levels played, and are still playing, an important role as mediators between ESCOs and current as well as potential clients. Energy agencies have also taken on the role of carrying out energy efficiency measurement and verification.

Barriers

The main barriers to ESCOs in industry are the unwillingness of clients to engage in contracts with payback times longer than a few years, and the reluctance to use ESCOs when the core production process is affected.

A serious problem for ESCO projects is the need to measure and verify savings, which requires a relationship of trust between the ESCO and the client. The client's willingness to cooperate with the ESCO is essential. To overcome some of these barriers, the Wuppertal Institute and its partners have developed a concept for a German "Energy Saving Fund."

One suggestion is to establish a guarantee scheme for ESCOs to overcome problems of insolvency of ESCO clients, which have increased in recent years.

In the public sector certain legal conditions (budgetary and municipal law) could be improved, as they hamper the work of ESCOs today. A neutral stance on how remuneration from savings should be accounted for within the municipal budget is one essential point which needs a clear definition. Energy efficiency related public contracts are usually simply awarded to the lowest bidder (up-front investment), and energy saving are not considered as a portion of life-cycle costs. The former ESCO association, PECU, therefore, requested that life-cycle costs of new equipment be taken into account in the public bidding process and that it become more transparent. Purely project based financing for performance contracting projects is believed to have the capacity to improve market uptake, but has not yet been in use. Several larger ESCOs are reluctant to bid for contracting projects in the public sector due to two factors: 1) the tender specifications are often considered as being of low quality and unclear, and/or 2) because of the small size of tenders and long and costly acquisition process.

Delivery contracting has come to a halt because of legal uncertainties due to government actions, and this is considered a step back for ESCO projects in the residential sector. According to a recent decision by the German Federal Court of Justice, the costs for investments in such a project can only be imposed on the tenant's costs if this was stipulated originally in the rental agreement, or if all tenants agree to the investment.

Nevertheless, the ESCO market in Germany continues to grow, with special increases in certain sectors, such as the hospital sector or industry, which are projected to grow by as much as 100-150 percent.

ITALY

The Italian energy service industry has been active for over 20 years. In the past, the ESCO market was stable, but not particularly large. The sector has changed in the last two to three years, when it was boosted as a result of governmental policies and as a consequence of market liberalization. The number of companies really offering ESCO services is not more than a few dozen companies, although the ESCO market has been increasing recently. The market is still dominated by large ESCOs, but small companies also have some ESCO services. Market size estimations vary widely. Some experts estimate it to be €60 million, based on the average annual turnover of the companies associated with Associanziaone Nazionale Societi Servizi Energetici (ASSOESCo), an ESCO association founded in 2005. This estimation is deemed rather conservative because members of the association are small ventures. Others estimate the micro-CHP market, where ESCOs are active, to be about €300-500 million. There are others, however, who put this figure as low as €160 million.

In the early 1980s, the first EPCs provided heat service to the public sector under chauffage-type contracts, sometimes using TPF. Cogeneration plants were commonly set up in hospitals two decades ago. Cogeneration in hospitals has been regarded as one of the most important targets of ESCO investment ever since, because of the high saving potentials due to cooling. In 2006 only, 80 MW of CHP were installed in Italy through ESCOs, which required about €95 million investment.

Italian ESCOs have developed from diverse origins, including "ad hoc" independent companies, equipment suppliers, fuel and/or electricity suppliers, public energy agencies, PPP and joint ventures, and ESCOs of French origin. The recent increase in ESCO firms is due to the entrance of individual professionals and small specialized enterprises.

Historically, ESCOs have tended to operate in the public sector. Other sectors have moved into focus recently, such as the commercial sector and industry. The residential sector is also getting attention, where boiler upgrading, heat control measures as well as small district heating installations for newly built dwellings are being carried out by ESCOs.

Government Activity

The current growth of the ESCO industry can be attributed to a complex set of legislative actions, changing market environment and international pressure. One of the most important changes in the regulatory background is the obligation for gas and electricity distribution utilities to reach end-use energy saving targets, this policy is also known as white certificates. Trading is encouraged in order to reduce costs of energy conservation measures, while penalties for non-compliance have been envisaged, too.

ESCOs are allowed to get the savings resulting from their project to be certified and sold. This setting increases the market potential available for ESCOs.

Registration of ESCOs by a government agency, AEEG, started in November 2004. Registered ESCOs are eligible for Energy Efficiency Certificates (White Certificates). The Italian regulation has adopted a very loose definition of ESCOs, which does not correspond to the stricter definition of ESCO used by the authors as defined in Chapter 1. The number of registered ESCOs companies cannot be used, therefore, as an indicator of the number of true ESCOs operating in Italy.

Energy saving measures implemented by ESCOs must be certified by the Market Operator, which issues certificates at the request of the regulator, AEEG, after verification. White Certificates acquired by ESCOs can be sold to distributors, who can cover their end-use energy conservation obligations. Some ESCOs attribute much of the increase of the ESCO market to the introduction of the White Certificate scheme in January 2005. The energy efficiency policy mix that has been advantageous for the EPC market has been complemented by the adoption of new building codes.

The Italian ESCOs often provide the financing themselves. Commercial banks are still scarce and over-cautious about financing ESCO-projects, and ESCOs have reported that only projects with especially beneficial parameters pass the banking criteria.

Barriers

One of the most significant obstacles is the loss of credibility of some participants of the ESCO market. On the one hand, as already described above, hundreds of companies have been claiming to be ESCOs because the accreditation was based on self-evaluation until recently. On the other hand, ESCOs do not trust clients because some industries and commercial clients may disappear during the contractual period (due to bankruptcy, translocation, change of activity, etc.). This situation is coupled with some uncertainty about the future legal environment. Moreover, the private sector is skeptical about the ESCO concept, and is suspicious about their own benefit from such an ESCO deal.

A major drawback to ESCOs in the public sector is that public sector regulations are not suitable for EPC. Tenders are traditionally price-based (based on initial investment cost) and energy performance (life-cycle costs) does not form the primary decision basis. Chauffage-type contracts are preferred. In this case, however, no energy saving guarantee *per se* is given and the savings are not monitored, although savings are normally realized.

The lack of interest from FIs has been a significant barrier. So far, ESCOs have mostly implemented projects using their own financial base. This, however, limits ESCO growth and the size of the market. ESCO growth depends upon banks becoming informed, and then participating in third-party financing.

An innovative suggestion has been drawn up by an Italian branch of an international ESCO to carry out a successful energy saving measure in a bank building. Since the central issue of their project was mutual trust and good understanding between the client and the contractor, they suggested that every working relationship be built up through a small-scale project,. The European Commission's Green Light Programme, which could serve as an introduction to further common business was put forward.

New positive policy developments include building certification, included in the national transposition of the Energy Performance of Buildings Directive 2002/91/EC, whereby ESCOs are authorized to issue buildings certificates. A new revolving fund with €25 million has been created to help the starting of TPF. Last, but not least, the Italian NEEAP has put a lot of emphasis on the role and development of the ESCO industry in Italy.

GREECE

The EPC market in Greece has not moved forward in recent years, and is still considered to be in its infancy. While there is the existence of a large energy conservation potential, principally in the services and the industrial sectors, ESCO activity is still negligible, and ESCO business in Greece has been restricted to only a few pilot EPC projects. Three companies have attempted to act as ESCOs in the past, but energy performance contracting has yet to be deployed, either in the public or the private sector.

The sporadic EPC projects to date have focused on renewable energy technologies (mainly solar thermal systems and small hydro investments).

Some upgrades have been done in lighting systems and in air conditioning. The ESCO-type projects in the past were commissioned by the government and concerned governmental buildings.

Barriers

The lack of ESCO business is blamed on the absence of a positive legal and institutional environment for the initiation and viability of ESCO operation. This includes the lack of clear, straightforward and supportive procurement procedures, and the absence of contractual and administrative guidance for the selection, control and repayment of energy services.

Licenses for power generation from alternative sources have been issued since 2006, which is expected to boost ESCOs active in renewables and combined heating and power (CHP). Nevertheless, the licensing procedure is not sufficiently streamlined and thus time-consuming, which is hampering fast and large-scale uptake of RES power generation. In 2005, only 3.4 percent of the total electricity generation was produced in CHP units. At the same time, support schemes have been introduced for CHP and RES, such as investment subsidies, leasing schemes, tax reductions and feed-in tariffs.

Financial and Government Support

Considerable interest in financing energy efficiency and ESCO projects is present on the side of commercial banks, and similar institutions, such as insurance companies. However, specific financial schemes, and procedures have not been developed as the system is not yet active.

The government has acknowledged this contradictory situation, and has recognized the opportunities offered by EPC; therefore, capacity building has been started, pilot actions have been initiated, and legal formulas (such as a law on TPF) have been drawn up.

Legislative changes have taken place, which are expected to foster ESCO activity. Law 3389 on Public Private Partnerships (PPPs) is expected to help the public sector overcome one of the long-standing barriers. Until recently it was prohibited to employ a private body to operate and manage the building energy services infrastructure of public establishments. The new law allows multi-year concession contracting for the installation, operation and maintenance of energy efficient equipment in buildings.

UNITED KINGDOM

Mr. Anees Iqbal, drawing on his long and extensive experience in performance contracting, has provided the following history and current perspective the ESCO industry in the UK.

The growth of energy performance contracting which is better known by the term contract energy management (CEM) in the UK, originated just prior to the first oil price shock of the early 1970s. It focused initially not so much on efficient use of energy as it did in providing cost savings. The term ESCO was not known in the UK at the time and the concept in the present form evolved much later.

The first companies offering ESCO-type services in the UK were essentially boiler house operating and management companies initiated by the National Coal Board, under the chairmanship of the Liberal peer Lord Derek Ezra. One of the first such companies owned by the National Coal Board was Associated Heat Service (AHS). Many other similar companies came on the scene and they all offered very similar services of providing operation and manning (using "Milk Round" concept) of central coal-fired boiler houses. The services provided were entirely supply-side energy management, and savings to the clients were largely generated by a reduction in operating and manning costs. AHS was later acquired by the French Company Générale de Chauffe, which itself was a subsidiary of Company Générale des Eau.

It was not until 1984 when Shell UK launched its own contract energy management subsidiary EMSTAR (Energy Management Services Technology and Resources) that true demand-side energy management service came upon the horizon. Prior to 1984, Shell UK had been trying out this concept from a small unit within the organization called Heating Management Service. Under this, Shell invested its own capital and technical expertise to modernize the heating plant in multi-dwelling buildings, installing new modular boilers and controls, clearing up the backlog of maintenance, and changing the fuel which very often involved converting oil-fired boilers to gas. This essentially proved to the market that the concept was not another vehicle of the oil company to sell more oil. In addition, the clients were offered guaranteed savings in their fuel bill. This activity proved highly successful in the marketplace and when the business concept generated significant revenue turnover, the business was hived out and the subsidiary EMSTAR was formally launched. Although this business was new in the UK, Shell, as a company had similar businesses in other parts of the world, notably in Holland and the USA. The business model for EMSTAR was largely based on Shell's activity already operating in the US under the name Scallop Thermal Management Inc. (STM). STM had concluded a major hospital demonstration contract in Philadelphia, which at the time was its flagship project.

Very soon after the launch of EMSTAR, British Petroleum also launched its own almost identical activity under the name of BP Energy. Thus the demand-side energy management business, until recently, has been dominated by the two oil giants in the UK—Shell and British Petroleum.

BP recently sold its CEM interests to ELYO Industries. Shell UK is also no longer in the CEM business. The French multi-national Company Generale des Eau (later named Vivendi) and Generale de Chauffe (later named Dalkia) in 1998 acquired EMSTAR (Shell UK's CEM subsidiary), and merged it with its UK CEM operation AHS, to create initially, AHS Emstar, and later, the name was brought in line with the Groups' international brand and named Dalkia plc in the UK.

The merged company became a wide spectrum ESCO providing both supply and demand-side energy management, and is now one of the largest if not the largest players in the UK CEM business. In the near future further mergers and acquisitions are apt to occur and some large group(s) may emerge to rival Dalkia.

Barriers

Initially, the concept of contract energy management flourished, in the UK, in the private sector. There were a number of barriers, however, in its path to be accepted in the public sector. The HM Treasury initially labeled CEM as an unconventional financing method and as such saw it as a back-door means of circumventing government's capital restrictions in the public sector. It was even labeled illegal for use in the National Health Service, as it seemed to violate a major clause in the National Health Service Act of 1977.

The CEM industry fought a long and hard battle with the HM Government Treasury, which ultimately saw the merit of CEM and permitted its use subject to detailed option appraisals demonstrating the "value for money" criteria. Today, of course, the Treasury not only approves the use of CEM but encourages all public sector bodies to consider private finance under its Private Financing Initiative (PFI) before approaching the state to fund major capital programs.

ESCO/CEM, however, still has a "rip off" image which stems from the apparent similarity with energy tariff consultant's style of contracts; i.e., negotiate lower utility rates and make a killing. Also, the attitude to PFI is very negative in the press and on TV. Observers generally decry PFI as very expensive for the public sector. Whether this is right or wrong, the negative publicity degrades what could be achieved through private finance under the ESCO/CEM heading.

Other barriers in the public sector include competitive quantitative tendering. While the supply-side energy services were relatively easy to evaluate on a competitive basis by comparing the cost of steam, the wider demand-side services were impossible to offer for. Novel ways of tendering such as "qualitative tendering" are being promoted by the industry to overcome this difficulty.

The lack of common contract terms is also a barrier. It makes tendering for smaller and medium-sized projects very costly and time consuming, taking up to 18-24 months.

Even though the ESCO concept is now a mature concept and has been used in the UK for more than 25 years, there is still an important lack of awareness and knowledge in the market-place about the concept and how to use it to extract the maximum benefit.

Types of ESCO Projects

The UK ESCO projects may be classed under three main categories:

1. **Demand-side refurbishment/retrofits** comprise, in addition to finance and performance guarantees, projects related to building envelope and hot water distribution improvements, insulation, controls, efficient lighting, boiler decentralization, energy recovery and routine and breakdown maintenance as well as fuel purchase and management.

2. **Supply-side retrofit/refurbishment** include activities such as boiler house retrofits; fuel switching; improved hot water and steam distribution systems, controls and insulation; medium scale CHP; fuel purchase and management; routine and breakdown maintenance; and finance and performance guarantees. There is also lots of scope for better skilled operation and management; i.e., a contract to run a facility without necessarily investing in new hardware. The potential for savings will be more clearly demonstrated in the public sector

when the Display Energy Certificates (DECs) program, initiated on 1 October 2008, proves itself. There is likely to be considerably more scope for achieving savings from better management, operation and maintenance of sites than currently achieved.

3. **New buildings** is a new area for larger ESCOs and those in consortia with M&E/FM contractors. This new business for ESCOs came about with the launch of UK Government's Private Finance Initiative (PFI) for Public Estate. This category includes provision of construction finance, turnkey contracting, operations and maintenance, total facilities management where required (catering, gardening, decoration, etc.).

Trade Association

As such, there is no association of energy service companies in the UK. However, the industry in the UK is represented as a CEM subgroup within the Energy Services and Technology Association (formerly Energy Systems Trade Association). The CEM subgroup was formed in 1987. The association encourages orderly growth of the industry through accreditation, support and advice to both CEM companies and customers. Current ESTA Memberships include equipment suppliers, utilities, and consultants in addition to CEM companies. Membership stands at 111 companies of which the CEM subgroup members total 11.

The ESCO Industry in the UK Today

The 11 members of the ESTA's CEM subgroup are now the major ESCOs in the UK. However, a number of smaller companies, who might call themselves ESCOs, also exist, serving mainly the smaller end of the market; e.g., single commercial premises, private dwellings, etc. Most of these smaller end clients will have an annual fuel bill well below £50,000 per annum. These new smaller CEM companies are usually a form of consultants which may have managed to acquire some financing to help their clients implement their recommendations. However, they may not provide the full breadth of service such as guarantees, and/or long term operation and maintenance. On the other hand, some O&M companies have also extended their services to include aspects of a traditional ESCO (such as access to some finance and performance guarantees) and may call themselves ESCOs. Some of these smaller ESCOs are now being taken over by larger companies, a notable example being Heatsave, a small company operating mainly in the London area, being merged with Cofathec, the Gaz de France subsidiary in the UK. This trend of mergers and acquisitions is likely to continue and further changes may take place in the coming months and years. Elyo (industrial/large plant market; parent Suez) and Cofathec are expected to merge this year.

Financing and Contract Models

In the UK, there is no single contract model in exclusive use. Shared savings, guaranteed savings, and the heat service (chauffage) models are all used depending upon the type of client and the type of project.

UK ESCOs use a wide variety of methods to finance their investments. Most ESCOs have access to pools of capital from private sector lenders or from their own organizations. ESCOs are able to structure the loan to be on, or off, balance sheet, depending on the tax situation and other considerations of the client.

Companies, such as Dalkia, frequently use the technique of undisclosed third-party financing. Under this arrangement, the ESCO has an agreement with a preferred bank or a finance house, whereby a separate contract for financing is not signed between the customer and the bank, rather it is implicit in the ESCO agreement, and the financier remains in the background. It is only in the event of the client defaulting that the financier steps in.

Major Market Segments

Unlike some other countries, the UK CEM industry has evolved in a very different format. The public sector was not the first market sector to embrace the CEM concept. As mentioned earlier, the Treasury placed many restrictions and the industry has had to fight a long and hard battle to get CEM accepted in the public sector.

Only recently has the public sector gotten properly involved with the industry through the UK government's private finance initiative (PFI). This has continued in many ways so that a private-public "partnership" has developed. Under these PFI partnerships, a significant number of new build projects are likely to be executed in the coming months and years. One such high profile project is likely to be the construction of the 2012 London Olympic Village.

The health and hotel sectors have traditionally been a lucrative market for the CEM industry. Both feed and sleep people in large numbers and the 24/7 demands for reliable energy services together with shortage of investment capital lends itself ideally to the CEM business model. This is a sector where the CEM industry has found a natural fit.

The same can be said for the multi-dwelling residential sector, early CEM contracts in the UK were largely in this sector. Education and commercial establishments are also sectors where the CEM industry has been highly successful.

The industrial sector is also a lucrative market for the CEM industry. This sector can be broken down into three main parts:
- The demand for reliable, and cost effective heat, power and compressed air;
- The space heating and hot water demands for the industrial building envelope; and
- Industrial processes.

The CEM industry has been admirably successful in the first two but has by and large shied away from getting involved in the processes of their industrial clients. CEM companies in the UK are also wary of direct involvement in social housing; i.e., where they have to get tenants to pay. Generally, they require guarantees from the local authorities or housing associations.

Enabling Factors

The UK ESCO industry has unfortunately not enjoyed the same strong support from their government as have many other countries. The UK government is a strong believer in letting the " market forces do the talking." However, as stated earlier, the government now does encourage the public sector clientele to consider the use of the private finance route as a priority over state finance provided that it offers a better value for the money. The UK ESCOs have had to fight hard to win large public sector contracts.

The credit rating of the public sector, as a client backed by the state, makes financing of their projects easy. However, the public sector insistence on competitive tendering remains a problem for demand-side management projects. Open book-type contracts prevail in the UK ESCO industry. This type of contract ensures greater trust between the ESCO and the client.

Factors Affecting ESCO Development

The opening of the public sector market and the significant increase in energy costs are certainly huge incentives for the development of the energy efficiency market, and therefore of the ESCO market. It is no longer a new or novel concept, and the fact that the ESCO market has now matured will certainly play an important role in its further growth.

CASE STUDIES

There are a number of interesting case studies available to demonstrate the wide range of projects implemented under the ESCO concepts in the UK. Most of these case studies are made available in open literature. Below are examples of such projects:

Project Title	Outline of projects
Industrial CHP	**Client:** Tunnel Refineries.
ESCO:	Dalkia Technical Services Plc.
Measures:	The client is a manufacturer of sugar syrups for the soft drinks and food industry. It is a large user of heat and power and its demand was set to increase due to expansion. (12-14 MWe Electricity and 40-50 Tonnes of steam) CHP was found to be the best solution by the ESCO who designed, installed and operates the CHP plant. The ESCO invested over £8.0 million in two new Tornado gas turbines and refurbished an existing steam turbine generator, under a CEM contract of 16 months installation and a 10 year management phase. Over a £5.0 per annum utility bill, the client enjoys £700,000 savings per annum.
Contract:	10-year (fixed and variable charge basis).
Project Investment: Total £8.0 million (1990).	
Annual Savings (energy and operation costs) £700,000 per year.	
Emission reduction: Not known.	
Hotel & Leisure Sector	**Client:** Devere Hotels.
ESCO:	Dalkia Energy and Technical Services.
Measures:	Devere Hotels had a portfolio of 34 (now much fewer) hotels spread all over the UK. The heating and hot water equipment installed at the majority of hotels was old, with poor controls and questionable reliability.

(Continued)

Project Title	Outline of projects
	Under a rolling programme, the CEM company invested its own capital and refurbished the heating and hot water plant, installed new controls and electronic BMSystems at each site. As part of the contract, the CEM company provides ongoing operation, monitoring and maintenance at all the sites, and provides regular management summaries to the hotel management.

Type of Contract: not available.

Approximate Investment: £1.5 million.

Annual Savings: Approx. 5 percent of the historic energy bill of the Group.

Emission Reduction: Not specified.

Industry

Client:	McVitie Biscuits.
ESCO:	Inenco Group.
Measures:	Inenco invested £1.0 million in a gas turbine based CHP plant and other energy conservation options as part of a CEM package. The project was executed as joint venture with McVitie. Inenco provided the engineering, design and installation which included a comprehensive building energy management system. McVitie benefited not only in energy savings but also enjoyed the replacement of outdated equipment with new technology and compliance with environmental legislation.

Type of Contract: Joint venture.

Projects Costs: £1.0 million.

Annual Savings: In excess of £3.0 million and significant CO_2 reduction.

Emission reduction: Exact figures not available.

The author would like to express his thanks to Mr. Alan Aldridge of ESTA for reviewing this article and making several helpful comments

IRELAND

The Irish energy services industry is still in its infancy. The ESCO sector is still underdeveloped. An early effort was commissioned by Sustainable Energy Ireland (SEI) to assess the potential for energy service companies in the country as a means to catch energy efficiency improvement opportunities

In 2005, 11 companies were identified that could be classified as energy service providers (ESPs), and two multinational companies were found to offer guarantee on their services in the form of EPC. The most typical (but still rare) motivation for potential clients is to outsource energy management to a specialized company, with or without the actual ESCO service and concept. The most prevalent contract model in Ireland is the BOOT model. On the other hand, Irish ESP companies do not often use EPC contracts, but prefer to work for a fixed service fee, and thus facing less risk. This is not primarily due to the reluctance on the part of ESCOs to engage in financing, but rather the disinterest of the clients.

Irish ESCO-type companies can be categorized in three groups:
1. Companies offering facility management, which comprises the management of the client's water and energy use, cleaning, etc.;
2. Companies offering contract energy management (CEM); or
3. Companies constructing and operating CHP.

Today, the estimate for the potential ESCO industry market size until 2020, is between €50-110 million/year. This calculation takes into account the 20 percent reduction potential of energy use in the EU, but considers hidden and missing costs, thereby reducing the potential. The authors of the report, *Assessment of the Potential for ESCOs in Ireland*, have applied various calculation methods in order to confirm the accuracy and came to a similar results.

Electricity market liberalization was completed in 2005 and gas market liberalization will follow soon. With restructuring, efficient cogenerated electricity is favored in the market. This is important for the development of ESCOs, as CHP is one of the most attractive areas for ESCO involvement in Ireland. In addition, investment funds at the Irish Energy Center under the Energy Efficiency Investment Support Scheme have been established, and the government has earmarked €5 million for CHP and district heating programs.

The most important market sector for ESCOs in Ireland is probably the industrial sector. The companies that are (at least somewhat) involved in ESCO-type activities have reported that industry accounts for 50-80 percent of their business, the commercial sector for 10-30 percent and the public sector for 10-20 percent.

Barriers

The main barriers listed by informants to the survey conducted by ENVIROS in 2005 included:
- the lack of governmental regulations and targets;
- a reluctance to outsource energy services partially because of concerns about redundancies in staff; and
- reluctance of potential ESCOs to take the risk of guaranteeing savings.

Furthermore, most of the potential customers are not aware of the ESCO concept. While the EPC concept is well known by companies that have the capacity to become ESCOs, lack of appropriate expertise at banks, high transaction costs and the lengthy contractual arrangements still pose an obstacle to higher uptake of this market.

FRANCE

Jérôme Adnot, a full professor at the Ecole des Mines-Paristech, and Frédéric Rosenstein, an engineer of the DSM division at ADEME, the French environmental and energy management agency, offer us an update on the status of ESCOs in the country that originated the performance contracting concept over one hundred years ago. Of particular interest is the unique classification of firms offering energy services, which evolved from the early combined HVAC operation and maintenance contracts.

A Short History of the ESCO Industry

Energy services (public lighting, gas and electricity distribution, district heating) in the form of outsourcing public services in France dates back into the 19th century. The success of these and other "delegated management" services (waste and water management, transport, telecommunication) financially strengthened the private companies involved in these businesses, thus creating the basis of the oldest French ESCO model. Traditionally, the "contract of operation" model dominates the French ESCO market. The French market cannot be fully associated with the definitions usually applied elsewhere. Originally it was based on the combined operation and maintenance contract of HVAC systems.

For HVAC system operations, the so called "chauffage contract" is a contract which includes operation without explicitly committing to carrying out energy efficiency investments. Under a chauffage contract, the contractor ensures optimal operation of an already existing system and must provide an agreed comfort level (for instance temperature, humidity) at a lower cost for the client. The contractor can increase its profits by investing in more energy saving equipment or by procuring cheaper fuel, thus reducing the costs. These types of contracts in France are usually long-term and include the obligation to diagnose problems and identify needs for improvement in the system, and a stimulus to carry out the investment, due to their dual nature (operation and installation companies).

Clients in the private sector applied the above contract type, but also became more flexible. The first formalized contract including third party financing (TPF) was signed in 1983. This was primarily designed for financing energy saving investments and to overcome clients' aversion to the high perceived risk of improvements that in reality were cost-effective, but not acknowledged as such by the clients. This model did not particularly spread in France due to the strength of the traditional "contract of operation" model..

The French operators have 'exported' the chauffage contract model to several other European countries, including Belgium, Italy, Spain, the UK, and Central-Eastern Europe. The basic concept of performance contracting, however; was brought from France by Royal Dutch Shell.

Due to the historical developments described above, traditionally clients of ESCOs were from the tertiary sector, and later from industry. In recent years, experts report increasing focus on industrial and residential projects, while the public sector is still the primary client of facility management contracts.

Although the total number of companies offering chauffage or EPC contracts is around 500, the French market is characterized by a strong concentration of actors, with only three large ESCOs dominating the market. These companies are subsidiaries of main energy utilities, though working independently from them. Earlier they were referred to as "expolitant de chauffage" while they now call themselves energy efficiency service companies. Recently, new actors have been entering the market. The new actors have different roots, such as big installers who provide financing in addition to traditional HVAC services.

Financing and Contracts

The most frequent ESCO contracts are still the first generation of contracts, based on operations contracting, with extensions.

Operating contracts involve heating and air conditioning installations for which the service provider has a firm commitment; undertaking, for example, a temperature level to be guaranteed for the heating of premises. The service provider is responsible for supplying the resources, which he considers necessary in order to achieve a specific result.

Contracts may cover the following areas:

- *P1*—Purchase of fuel, cost of heat energy (oil, gas, etc.)
- *P2*—Daily operation, cost of labour and minor maintenance
- *P3*—Full maintenance, cost of major maintenance and of the total guarantee
- *P4*—New equipment funding, investment depreciation

The law specifies a framework for the duration of heating and air conditioning contracts in the public sector. A central commission for contracts defines the main types of heating operation for public contracts through guidelines published by the Ministry of Energy and Industry: "Guide de rédaction des clauses techniques des marchés publics d'exploitation de chauffage avec ou sans gros entretien des matériels et avec obligation de résultat." Contracts which include P3 services have a maximum duration of 16 years. For contracts without P3 services, the maximum term is eight years in the case of fixed-price type contracts and five years for all other contracts.

P1 contracts: supply of energy, without explicit incentives. It is generally the largest expenditure item in a contract, except for the case of a large-scale multi-service P2 arrangement discussed below. This item may correspond to a fixed price or be proportional to the quantities of heat or energy delivered; it may also provide an increased profit if consumption is reduced.

P1 contracts with explicit incentives. An incentive clause can be incorporated into any of the types of contracts. The clause provides for the

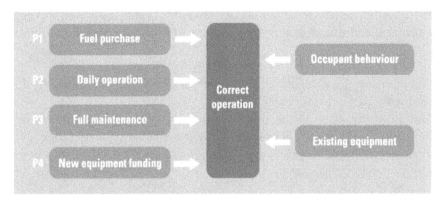

Figure 2-1. Practical Breakdown of Service Contracts

sharing of energy savings achieved or of excessive consumption, relative to a defined base consumption for a given heating year. This quantity is adjusted according to the actual period and weather for the given heating season. The incentive clause cannot be applied for the first heating year if the installation is a new one. The advantage of these incentive arrangements is that the occupant and the service provider are both encouraged to make energy savings, because they share the profit made. It should be noted that incentive formulas are defined by the central commission for contracts in the Schedules of General Technical Specifications (Public Contracts Collection no. 2008).

P2 contracts: Control and routine maintenance of installations. The provision of labour and consulting is the basis of all operating contracts. It involves the control and routine maintenance of installations. The contract may also include P1 services, representing the supply of energy, and P3 services for the major maintenance and replacement of installations, known under the term "full guarantee."

P2 type contracts, therefore, offer little direct potential in terms of energy efficiency. They provide for the correct functioning of installations over a period of time, with constant adjustment of combustion and efficiency settings. The operator is not, however, concerned with the supply of energy; his sole obligation is the correct functioning of the installations, which is not the subject of precise measurements of parameters such as efficiency. The obligation to maintain, clean, balance and inspect equipment is nevertheless a contribution to the improved energy efficiency of installations. An operator wants, however, to keep a contract and therefore develops a range of proposals and dispositions designed to satisfy the client, together with an offer for ancillary works, all of which may have an energy efficiency content and contribute to "constant progress." This type of contract is also known as an "'PF'" contract (fixed-price services).

P2 contracts with incentives. Although a P2 is not a contract for the supply of energy, it is possible to define an energy saving incentive in a pure P2, which is equivalent to buying a service for the continuous improvement of operation.

P3 type contracts: Major maintenance and renewal of equipment. This involves the major maintenance of installations as well as any equipment replacement. In contracts of this type, the building owner pays a fixed annual fee depending on the age and condition of the installations. In exchange for this, the operator agrees to replace all or part of any defective equipment during the term of the contract. Although it is transparent, a majority of public-sector buyers see it as a purchase on credit, which is not appropriate.

Under P2/P3 contracts, the operator has total responsibility for the installations and can elect to replace equipment with a view to reducing operating and functional costs. The operator is contractually obliged to replace installations with identical equipment and to provide equivalent performance. This type of contract, which seems to correspond to the budgeting logic of the state, cannot however be applied in the context of state contracts because it makes use of anticipated payment procedures which are prohibited in public contracts.

This type of arrangement, therefore, has a significant energy efficiency content if it is entered into in a P1 type contract with an incentive clause. If there is a joint agreement with the owner to replace existing installations with installations of higher combustion efficiency, the cost difference can be funded by using the P4 clause, which offers a particular mode of funding over a period of time.

New ESCOs come mostly from the manufacturing world (Siemens, Johnson Control, Schneider Electric, etc.). First of all, the ESCO offers to carry out a free summary audit on the client's site. This audit identifies the directions in which progress can be made and provides an order of magnitude for the investments to be undertaken, the energy savings to be expected and the payback time.

Current ESCOs Industry and Market

Part of the ESCOs belong to an association, Fédération Française des Entreprises Gestionnaires de services aux Equipements, à l'Energie et à l'Environnement (FG3E), which has around 500 members. The annual turnover of this market is estimated by FG3E, to be €3 billion. It is impossible to know which part of that market includes a share of real performance contracting.

FG3E has published statistics on operating contracts:
- 115,000 installations are covered by operating contracts
- 40,000 of these contracts include P1 services, two thirds are at fixed prices and the remaining third corresponds to contracts with metering
- only 10 to 15% include an incentive clause
- estimated final energy consumption under P1 contracts = 33.3 TWh/year, including:
  — natural gas: 20 TWh/year (60%)
  — oil: 9.9 TWh/year (30%)
  — others (coal, wood, geothermal, etc.): 3.3 TWh/year (10%)

The companies involved in energy efficiency services formed an association formed in 2005 under the name "Association of Energy Efficiency Service Companies" (CS2E). This effort was instigated by ADEME in connection with the AIE (task X) work on the European "Best" and "Eurocontract" projects. The club members at present are FG3E, GIMELEC, UFE (Union of French Electricity Suppliers), UCF (French Climate Control Union, a group of installers in the field) and SERCE (Syndicate of Electrical Engineering Companies).

The aim of the association is to accelerate the development of the market for energy efficiency services, including energy performance contracts. The mission of the association is:
- to inform all those concerned on the following subjects:
  — potential sources of energy savings;
  — the possible savings which are technically and economically achievable;
  — the typology of offers, solutions and funding; and
  — the positioning of the players.
- to contribute to developing tools and regulations.

Services Provided

On the basis of the results of the audit, if they are satisfactory, the ESCO invites the client to sign a preliminary agreement or memorandum of understanding with the following conditions: A detailed audit will be carried out after the memorandum of understanding is signed. If the detailed audit confirms the result of the summary audit in terms of the value of energy savings with a certain margin of error (on the order of 10%), the client then has the choice between signing the energy performance contract or not signing it, in which case the client has to pay the ESCO the cost of the detailed audit. If the client signs the contract, the cost of the audit is built into the overall cost of the project.

Various contractual formulations can then be envisaged, all of which have in common a guarantee on the value of energy savings identified in the detailed audit. The ESCO's remuneration is based on the value of the energy savings achieved and, if these are lower than the guaranteed savings, the ESCO pays the difference.

The factors which differentiate between the contracts include the following:
- The share of the investment assumed by the ESCO, which can range from 0% to 100%.
- The duration of the contract, this depending on the value of the guaranteed energy savings, but also on the share of this amount. If this share is zero, the energy savings pay off the investment and remunerate the ESCO; if the share is non-zero, the duration of the contract increases.

There is a third generation of new companies, which are attempting to add some "smart" controls to the usual operational control. They act has consultants but accept to be paid from the savings. They don't take the place of the operational companies, but they request the right to set the control values from outside into the building energy management system (BEMS). ERGELIS, for example, proposes materials and services and remotely controls the main equipment consuming and possibly producing energy in the building (online or via the existing BEMS, through relays). Piloting is carried out by regularly sending of instructions to the relays through modems. The data-processing server acts as centralized point for the dedicated software of optimization, which simulates permanently all possible scenarios of control and determines "optimal" control to set up (based on consumption and invoice).

To assemble its offer the ESCO studies the histories of consumption, evaluates the economy and proposes to the customer a package which may include installation of a relay box connected to the systems of heating and air-conditioning; remote piloting, by fixed telephone network or GSM network, installation of a box connected to the systems of heating and remote piloting, by fixed telephone network or network GSM; installation of a supervision of energy, accessible by Internet to manage the schedules of occupation; relay allowing the management of the peak demand of power, obtained by reducing consumption for the peak periods to lower the contractual demand and while acting at real time on certain equipment (electric batteries, for example...) to avoid the excess charges; and dynamic management of the production of heat or cold, according to the parameters of use, prices, and weather.

An initial inventory of features of the operation in place is made. It is expressed in "customer-friendly" terms: ranges of temperatures, schedules of occupation, etc... It will be respected in the new operation. However the service is a "black box" for the owner and for the local operations company; the gains will disappear when the contract is ended.

A New Type of Public Expenditure

The recent appearance in France of public-private partnerships (PPP) is changing the investment funding context. These special contractual arrangements should in effect allow performance targets to be introduced into invitations to tender, particularly with regard to defining the level of energy consumption to be attained. PPP contracts are global administrative contracts by which a local authority or a public establishment associates a third party with the funding, design, production, conversion, operation or maintenance of public equipment, or with the funding and management of services. This type of contract is not a public contract or the delegation of a public service, but represents a new category of public contract that has a specific award procedure.

The technical and economic analyses required in order to use this procedure might dampen the enthusiasm of local authorities, who wish to take up this new tool to fund their investments in energy efficiency. The definition of a functional programme to open the dialogue with the future contractor requires total motivation on the part of the local authority that nevertheless wishes to have recourse to this instrument. In the final analysis, as the legislators have said several times, the partnership contract should remain an exceptional solution, destined mainly for new building projects. However, the Law changed in 2008, but all details are not known. Some experts envisage an explosion in the number of PPPs.

Types of ESCO Projects, Major Market Segments

Primary projects implemented by ESCOs are still HVAC system operations, public lighting, compressed air production and building modifications, CHP and facility management. French ESCOs mostly provide complex solutions.

Barriers to ESCOs and EPC

One of the most important legislative restrictions, which impedes complex ESCO activity in the public sector is the prohibition of operation and particularly purchase of equipment in the public sector being designated to private entities, except in the scope of very special and formal public-private-partnership (PPP) agreements.

It has long been claimed by ESCOs and the FG3E that the engagement of the private sector to provide complex solutions for the public sphere would be beneficial, and innovative solutions could appear as a result. Therefore a government order creates the possibility to draw up PPP contracts where a concession scheme is not available and where traditional procurement contracts (marchés publics) cannot be implemented because of the legal restriction to have separate contracts for each phase of the design, construction and operation of a project.

The new order also allows the public sector to pay the private company's remuneration periodically during the project, and allows that payment to be based on performance indicators previously set out in the contract (instead of being purely revenue based). French ESCOs see the opportunity to be able to use savings in operations budgets for investing in efficient equipment as important.

The private sector normally pays primary attention to its core business. It has been found in France that, without fiscal incentives, private companies and households do not engage in energy saving measures. Until recently, the price of energy has not been high enough to encourage savings in these sectors. The promotion of energy efficiency, and ESCOs as a tool for that purpose, would help. Furthermore, the social housing sector (and in general, rented houses) need special treatment to overcome split incentives.

Future Opportunities

A "White Certificates" scheme was introduced in France in 2006, and is expected to enhance energy efficiency services in the private sector when coupled with the recently rising energy prices. It is expected that the ESCO market will accelerate. Therefore, recent legal developments are expected to further boost the French ESCO industry.

CASE STUDY: STREET LIGHTING OF A CITY BY AN EPC

Context

In 2004, the city of Lille and the common partners of Hellemes and Lomme renewed their market of street lighting.

The invitation to tender street lighting is not isolated but results from a progressive reformulation of all the markets within the framework of Local Agenda 21. It answers the criteria that the municipal team defined for the making of new public markets:

— improvement of the service (equity, quality, access to all, etc.);
— control of the total cost, impact on the local development; and
— reduction of flows, control of ecological print.

Description of the Invitation to Tender

The invitation to tender "total maintenance and maintenance to certain level of the performance of the works of street lighting" included degrees of negotiation dealing all with sustainable development:

— energy savings;
— renewable energies;
— recycling; and
— other advantages for the citizens.

This invitation to tender was composed of four criteria:
— "energy": supply and energy management, remunerated by the monthly fixed price
— "maintenance": preventive and corrective interventions and follow-up of installations, remunerated by the monthly fixed price
— "breakdown service": interventions in reaction to unforeseeable events, annual remuneration adjusted compared to one working year of reference
— "rebuilding": revising, putting in conformity, putting back

in operation, repair, modernization and improvement of the equipment, annual remuneration according to operations carried out.

The total envelope of the call for tenders was fixed at €4.4 million (M€) per annum, that is to say 35.2 M€ over the 8 years of the contract. This contract relates to 22 000 luminous points.

The call for tenders constituted a negotiation "on performance": each tendering body was invited to propose a level of engagement, like describing the means to implement, the indices of control of the performance and the associated penalties (those having to be adapted to the level of performance suggested). This organisation made it possible for the municipality to get information about the level of performance that it was possible to reach.

Note: this set of tendering requests is available for other municipalities wishing to take a similar step.

Proposal Adopted

The engagement on the consumption of energy has been carefully examined. The consortium retained is ETDE/SOSIDEC. It is committed to saving 42 percent of the power consumption.

Beyond 42 percent the benefits are shared (50-50), under this value the company pays penalties. In addition, the company began to partly feed the network of street lighting with electricity of renewable origin: 25.7 percent of electricity has been certified by Observ'er in 2005. The holder of the supply agreement of electricity is the company EDTE and not the community.

The savings generate a redistribution of the expenditure. In addition, this offer sets out the expenditure in the following way:
— energy: 19 percent (37.6 percent in the preceding contract)
— maintenance: 21.6 percent (32.5 percent in the preceding contract)
— rebuilding: 59.4 percent (29.9 percent in the preceding contract)

Assessment of the 1st Year of Operation

Reduction of the consumption of electricity and of the power call:
— drop in total cost of the service from 210€ (including all taxes) to 200 € (including all taxes) by luminous point, which is equivalent to a fall from 21 € to 20 € per capita.
— reduction from 5000 kVA to 3500 kVA; this is 30 percent of

economies without reduction of the service (2004: 20,6 GWh; 2005: 14,3 GWh) So over the first year 1.3 M€ were saved and transferred from the lot "operation" to the lot "investment."

Actions already performed include:
— reduction of over illumination: installation of ballast and power dividers;
— suppression of 1,048 luminaries of the "ball" type (luminous pollution); and
— management of waste: lamps, masts, consoles, 98% of the mass of waste. In addition, 15 jobs were created in the companies partners in charge of the management of this waste.

AUSTRIA

Austria is another success story of the ESCO industry in Europe, and the particularly fast uptake is an exemplary case for the rest of Europe. Austria offers numerous interesting case studies, which are highly replicable.

The ESCO market in Austria had a rather late start. The level of the ESCO market was nearly zero in 1998; followed by a quick take-off in less than a decade. Austria has become an ESCO market leader in Europe. As of 2006, there were around 30 ESCOs in Austria, and the number is still increasing, though only five companies cover 70-80 percent of the total market. ESCOs estimate there is €500 million investment opportunity in economically feasible projects for the rationalization of energy use.

The general financing scheme in Austria has been the shared savings model. Bundling of similar projects, following the example of Berlin, has proven to be an important success factor. Increasing and guaranteeing the quality of projects is a priority, and for this reason standard documents (such as contract models) have been made available, and standardized project development has been introduced. Uniquely, even among the developed ESCO industries, several quality labels have been set up for ESCOs and ESCO services. The Thermoprofit quality label initiated by the Graz Energy Agency was introduced to guarantee reliable high quality proposals by ESCOs using the label. The label is issued by Graz Energy Agency and an independent commission that assesses the ESCO companies at regular intervals to confirm that they fit Thermoprofit standards and this example has spread to other regions. The so called eco-label, on the other hand, denotes the quality of ESCO services and the compliance with standards.

The great majority of the EPC contracts until now have been concluded in the public sector, in federal and municipal buildings. The private sector is lagging behind. Between 1997 and 2005 over 1000 public buildings were optimized with the EPC tool. In 2004-2005 another huge federal program started with about 800 buildings. On average, ESCOs have been able to guarantee almost 20 percent savings for 10 years in these contracts. Improvements have been achieved on heating and cooling systems, lighting, and water management. Street lighting has been renovated widely, too. There are dedicated programs to increase energy efficiency in municipalities, such as the e5 programme under the national climate protection program.

In recent years, more and more effort is being given to increase the number of ESCO projects in the private service sector and to find out the reasons for the slow uptake of the ESCO model, in spite of the same or higher energy saving potentials as in the public buildings. Various programs have tried to find and remove barriers in this sector. It has become clear that barriers are larger in the private sector both on the clients' and the contractors' side.

The building owners and/or users still lack awareness about the benefits of energy efficiency and the opportunities offered by ESCOs, even though energy related costs constitute up to 50 percent of the operating costs in private service buildings. Private buildings are often rented out, creating classical split incentives. Furthermore, energy related matters are seen as less important compared to core issues, and consequently private companies pay less attention to this area. It is perceived that decreasing energy demand does not add much to profitability. Finally, the private building owners are often hesitant to get involved in long-term contracts, and some are scared by previous bad experiences.

A limited number of projects have been implemented in shopping centers, hotels, banks, churches, office buildings, and hospitals.

Renewable energy sources have started to get attention, too, during the last few years. Currently there are 3 million m^2 of installed solar collectors in Austria. Graz (250,000 inhabitants) has an innovative district heating system, that integrates a 10,000 m^2 sized solar collector surface for supplying a 2500 MW thermal energy per year. This area is clearly growing, opening new fields for ESCOs.

The government has played a significant role in the sharp development of the ESCO sector in Austria. A number of incentives are available for investments for the rational use of energy (subsidies, soft loans, tax credits for residential buildings). The involvement of federal and municipal buildings to the extent described above is exemplary. Energy agencies have been very active, participating directly and indirectly in ESCO projects. Obligations have not been typical, but in a few regions audits are obligatory in public buildings. Finally, ensuring quality and developing certification of ESCOs and ESCO businesses are credited for industry growth.

BELGIUM

The ESCO business started in Belgium in 1990. Some initiatives were taken by the federal and Brussels regional governments to promote the concept of third-party investment.

The first projects where mainly done by large companies offering facility and building management. The chauffage model was used in the public sector; e.g., for the Ministry of Defense. New entrants were TPF-Econoler, a joint venture of the Canadian ESCO, Econoler, which started offering third-party financing for buildings; e.g., for the city of Charleroi, and FINES, a Belgian ESCO specialized in lighting projects.

The public sector (mainly sports halls and schools) has received attention from ESCOs in Belgium. The industrial sector was also targeted, particularly for supply contracts. Willingness to outsource by large consumers has been an important driving force, in order to provide off-balance sheet solutions for energy efficiency investments. EPC with performance guarantees did not really develop early on.

The market for ESCO services did not fully develop, until the creation, in 2005 by the federal government, of Fedesco, a public ESCO focused on energy saving projects in federal buildings, using third-party financing. Fedesco, a limited liability company (NV/SA) under public law, is a subsidiary of the Federal Participation and Investment Company, a government owned financial holding. It was started with a capital of €1.5 million from the Kyoto Fund, raised to 6.5 million in 2007. Since January 2007, Fedesco has an exclusive right to apply third-party financing to federal buildings. Fedesco manages turn-key energy services projects on behalf of the building occupants and in collaboration with the Federal Building Agency and in 2008 has started the development of energy performance contracting (EPC) projects. Fedesco is also developing a federal competence center on third-party financing (TPF) and EPC. In late 2008, the federal government approved Fedesco's objective of reducing CO_2 emissions by 22 percent by 2014. This represents a five year gross investment plan of €210 million. Roughly 45 percent of those investments are scheduled to be done through EPC.

Since 2007, there's been a growing interest for EPC, stimulated by Fedesco and growth opportunities from multinational companies in building automation and control including Siemens, Honeywell and Johnson Controls. They have realized a few very successful projects in hospitals and schools. Also large multinational companies, such as Dalkia, and Axima Services, which have been offering facility and building management, are now starting to offer EPC. The TPF group is also redeveloping EPC offers.

On the consumer side, the city of Antwerp has plans to start using EPC for its own municipal buildings. Technologies targeted in the public and industrial sectors are lighting renovations, improvement of heating and cooling systems and control systems. Dexia also offers PV, solar panels and cogeneration (CHP).

Financing of ESCO investments is not a problem, and it is not a factor that limits development. Customer financing, ESCO-based funding and TPF (mainly leasing) are all used in Belgium. ESCO-based funding is often preferred in order to limit participants and to have only one responsible partner for the entire project.

Some party investors emerged; e.g., Green Invest and one of Belgium's largest banks, Dexia. They developed a specific third-party financing offer, called Energy Line. Energy Line which has focused mainly focused on public authorities like cities. Dexia collaborates with specialized engineering companies for the technical know-how. Energy Line will most likely be extended to include other offerings, including full EPC projects.

The federal and regional governments have taken important steps towards increasing energy efficiency. Besides transposing and implementing EU legislation, other measures, such as voluntary agreements, energy saving obligations for utilities, green certificates and public sector obligations have been aimed at increasing energy conservation in Belgium.

Fedesco has been asked to stimulate the development of the ESCO market, on the demand side, through the creation of a federal competence center for third-party financing and energy services, and at the supply side through joint efforts with private ESCO's, third-party financing companies and other market players. The Belgian ESCO Association, BELESCO, started its activities in 2008 and expects to focus on representing the ESCO industry, information dissemination to, and training of, public and private customers, building a database of EPC projects, developing an accreditation program and creating a model contract and tendering procedure for the public sector.

The public and private sector is specifically targeted by companies like Enfinity and Ikaros Solar, who use private funds for solar contracting, based on third-party financing. Solar ESCOs install, operate and own the PV Solar projects, offering rental prices (per m^2 of roof space or installed KWp power) as well as "green" electricity at reduced prices. They are targeted at large industrial, installations and public buildings, mainly in Flanders, with roof surfaces larger than 1000 M2. Similar offerings exist for residential customers, mainly in Wallonia and Brussels, driven by green certificate schemes that are advantageous for smaller installations, typically 10 to 50 kWp.

A small part of the industry is currently initiating activities in the residential sector. It is the aim of smaller energy consultancies to complement auditing services with the sales and direct installation of energy efficient household equipment and lighting. Nevertheless, the residential sector is still a minor client for ESCOs. Another federal government's initiative is the Fund for the Reduction of the Global Energy Cost (FRGE), a fund of €100 million financed by a federal state obligation loan. Through local entities at the city level, households can get low interest loans for energy saving measures, with a full service package for a target group of low income households. Regional governments have recently created eco-renovation products in collaboration with banks.

ESCO Industry and Market

There are now roughly six large and some smaller ESCOs in Belgium. The ESCOs' specific activities are poorly documented, as they are mainly operated by private companies which are under private agreements, except in the case of public sector projects. BELESCO is expected to be able to provide more data in the near future.

Financing and Contracts

In Belgium, the guaranteed savings model is the most commonly used model. Most ESCOs have access to private capital from banks or from their own organizations and are able to structure the loan to be on, or off, balance sheet, depending on the tax situation and other considerations. Bank guarantees do, however, add to credit exposure, creating some issues for smaller ESCOs.

Fedesco started with a €5 million financing capacity, with a state guarantee, which was increased to €10 million in 2007. This will most likely be increased further to provide third-party financing of at least part of the €210 million investment plan. Fedesco uses an EPC contract based on the plan developed by the Berlin Energy Agency and other partners in the EURCONTRACT project. Tendering procedures and contracts have been specifically adapted to the Belgian public tendering law. It will be the basis of BELESCO's effort to develop a common model contract for the public sector.

Additionally, most utilities and regional energy efficiency agencies have a set of incentives targeting public, commercial, residential and industrial sectors, which bring partial financing for eligible projects. The federal government provides tax reductions on energy saving investments; 13 percent for private companies, 40 percent (limited to a certain amount) for residentials.

Types of Projects

The ESCO industry in Belgium has been very concentrated in the public sector, and in the commercial building sector. Projects include replacement of boilers and chillers, cogeneration, relighting and other technical measures. Fedesco also includes measures at the level of the building envelope like insulation or window films.

Overall, industry is more reluctant to outsource energy efficiency services although this is expected to change with time.

Barriers to ESCOs

One of the biggest barriers is the lack of awareness and knowledge about the concept and how to use it, both in the public and private building and industrial sector. BELESCO is expected to play an important role in that area.

Another barrier is the large number of existing long term maintenance contracts, including a growing number with total guarantees on technical equipment. It is difficult to replace these by EPC contracts.

In the public sector, there is an important number of facilities which may prefer alternative renovation projects with their own funds, managing technical measures individually or using traditional credit funding. At Fedesco, roughly half of its investment plan will not be based on EPC projects, but so-called transversal measures; e.g., for boiler or chiller replacement, building control, insulation, window films, etc., not requiring full energy service offerings or EPC contracts.

Enabling Factors

The Belgian ESCO industry has strong support from the federal government. Through Fedesco, the federal government grants ESCOs immediate access to large public sector contracts for its own 1,800 buildings. The credit rating of the government as a client makes financing projects easy and the industry is set to grow.

Energy efficiency programs are however, managed by regional governments. It is unclear what support they will give to the development of the ESCO industry. This will depend on the results of the transposition of the EU Services Directive in Brussels, Wallonia, and Flanders. In this sense, the obligation for distribution companies to implement and fund energy efficiency will likely bring about the emergence of new energy service offerings by public actors.

Future Expectations

It is expected that the ESCO market will grow significantly in Belgium in the next five years. Fedesco's five year investment plan including EPC and its federal competence center on third-party financing and energy services, as well as the creation of BELESCO will provide key impulses to the market.

Private ESCOs like Axima Services, Dalkia, Cegelec, Siemens, Honeywell, Johnson Controls, and third-party investors like Dexia, Triodos and Green Invest will further develop their EPC offerings.

The long-term trend of rising energy costs everywhere is certainly a huge incentive for the development of the EE market, and therefore of the ESCO market. The fact that the ESCO market is better structured today than in its early days in the 1990s, will certainly play an important role in market growth in the near future.

THE NETHERLANDS

The Netherlands has successfully promoted energy efficiency, but without significant energy service company activity. Energy management is common; however, there is almost no energy performance contracting. There are only a few ESCOs active in the Netherlands. The level of activity and the number of companies have not changed much in recent years. The market is small, although no exact market potential has been estimated.

In the case of the industrial sector, voluntary agreements have been successfully pushing energy efficiency improvements and industry has been implementing measures on its own, given that they possess the financial and technical means and in-house capacity. There has been, however, a market for specific consultancies to support these changes. Lately, voluntary agreements cover the tertiary building sector, too.

Energy efficiency improvements in the residential sector are supported by other means than ESCOs, including grants, and preferential loan rates. 80 percent of all rented houses are social houses, which are occupied by lower-income people at low rates. Improvements in the social housing could be potentially an important market for ESCOs; however, as a result of governmental programs, this sector is rather overcapitalized and ESCOs are not able to compete for projects.

In the case of the public sector, the role for ESCOs has been limited because there is one organization responsible for the management and operation of all state owned buildings. There is another one for military sites, which takes care of energy related investments and refurbishments on their own, and they are not interested in employing an ESCO. These organizations themselves stand close to the ESCO definition, but without guarantees and traditional risk-sharing. Nevertheless, implementation of energy system improvements is supported with energy performance calculations.

Street-lighting and large buildings that are not state owned (hospitals) do make up a segment that is available and open for ESCO contracting. Complex projects often take place through joint ventures.

Since the energy efficiency market has been moving without ESCOs, potential clients have not seen positive examples and do not count on this solution for energy saving. It is evident that the ESCO is only one of the tools for increasing energy efficiency, and the Netherlands has been using other measures to become one of the leaders of energy efficiency.

LUXEMBOURG

Specific information about the ESCO market in Luxembourg is scarce. The number of ESCOs present in Luxembourg is around four, which includes daughter companies of large multinational companies of French and German origin, but also one Luxembourgian company. Occasionally, ESCOs in the surrounding countries implement projects in the country.

Energy intensity in Luxembourg has decreased 2.5 times faster than the EU average between 1990-2004. To this end, lots of measures related to energy conservation and rational use of energy were introduced in order to support the achievement of Kyoto targets and other commitments. These measures include voluntary schemes with industry, with hospital associations and the banking sector, subsidies and fixed feed-in-tariffs for RES. Support also exists for households and the public sector to implement energy efficiency with a maximum subsidy of 40 percent of audits carried out for the buildings. Specific measures to support ESCOs, however, have not been central to Luxembourg's EE efforts.

FINLAND

The Finnish energy agency, Motiva, maintains a public list of ESCOs and an ESCO project database. While there were three ESCOs identified in 2003. In recent years the number of active ESCOs registered in the database of Motiva reached nine. Experts at Motiva are aware of a total of 11 companies that have implemented at least one ESCO project. Four to five of these companies are actually actively participating in the sector. One ESCO company has roughly 90 percent of the market share. There are six local ESCOs or subsidiaries of multinational companies, two local energy companies, and three other companies that have had several ESCO projects. The EPC business has been increasing lately, but not to the extent previously expected.

Industries are the focal point for energy efficiency investments by ESCOs. Energy intensive industries, such as the paper, chemical industries and metallurgy are increasingly making use of ESCO-offered services. These industries are interested in energy savings because 15-20 percent of their costs are energy costs. Thus, production processes and heat recovery have been the most common ESCO project areas.

The public sector has also been addressed widely. In regards to the numbers of projects, 50 percent have been carried out in this sector, though regarding the level of energy savings, the public sector accounts for only 10 percent of total savings resulting from ESCO projects. The public sector contracted ESCOs for HVAC system improvements in the most cases. Both shared savings and guaranteed savings contracting models are used in Finland.

No recent estimate of the size of the ESCO market in Finland is available. The latest information is for the period 1998-2004, when the annual savings through energy efficiency was roughly €95 million/year; 5 percent of which was carried out by ESCOs. ESCOs estimate are that as of 2004 not more than 10 percent of the constantly growing industrial ESCO market potential had been captured.

The most important and successful push for energy efficiency in general and for ESCO contracting has been the Voluntary Energy Conservation Agreements between the industry and the Ministry of Trade and Industry introduced as long ago as 1997. There is an on-going auditing program supported by the government; but this has not created a large increase of ESCO activity as clients generally implement the suggested measures themselves.

From a 2006 report summary on the voluntary agreement compiled by Motiva the following data about ESCO projects are offered:

— out of all investment projects 11 were ESCO projects and they received 26 percent (€1.74 million) of all investment subsidies;

— out of investment subsidies to industrial projects the ESCO projects received 37 percent in 2006;

— accumulative investment subsidies during 2002 - 2006 for industrial ESCO projects (27 projects, €3.7 million subsidy) were about one quarter of all subsidies for industrial projects; and

— total investment in industrial projects is estimated to be at the level of €24-31 million as usually the subsidy varies between 15-20 percent of investment.

Financing of energy efficiency investments has not been a problem in Finland, in spite of the limited activity by banks. Both clients and ESCOs have the capacity to obtain financing for the projects. An increase of awareness in the financial sector about ESCO industry would have the projected effect of boosting the market by increasing financial input and involving new ESCOs, clients and projects.

Other driving forces are the increasing energy prices and environmental requirements, limited in-house energy expertise in the industry and sometimes limited budgets for refurbishment.

Barriers

The normal procurement process does not recognize EPC. Standards for procuring ESCO services are needed. The new accounting system makes the bookkeeping of ESCO projects more complicated: according to the new IAS/IFRS reporting, equipment must be shown in the client's own balance sheet and investment budgets, and booked as a financial lease. This way, an ESCO service is booked in the accounting in 3 parts: services, lease and interest. This is unfavorable for the ESCOs, which offers a complete service package and not equipment separately. More importantly—because the new equipment appears as clients' investment—decisions must be taken by the client according to their internal investment rules. Earlier, the ESCO service could be handled simply in the income statement as a purchase of services.

Opportunities

Industries that have joined the voluntary agreement described above are eligible for 15-20 percent subsidy of the energy efficiency investment costs from the government. Furthermore, the subsidies are peculiar in Finland since they are designed to help the ESCO industry by offering an additional 5 percentage point subsidy to clients if an ESCO is employed in the project.

SWEDEN

The Swedish EPC market has been dormant until recently. A quick growth has been observed during the last two to three years, driven by the growing interest from potential customers. For a long time, a lack of experience, mistrust and legal ambiguities hindered the uptake of the benefits offered by ESCOs. Mistrust in ESCOs and EPC is a particularly important issue in Sweden. The mistrust developed because ESCO-type investments already took place as early as 1978. Unfortunately, many of these failed as they did not yield the anticipated savings result.

The ESCOs and the Market

By 2007 the number of ESCOs offering EPC was around 12-15, an increase from around 5 in just the previous two years when two larger consultancy firms and one HVAC company started to develop ESCO-like services for small-scale projects. Now there are local market actors, control companies, building service companies and consultancy companies. Some of the EPC providers have extended their structure and formed special EPC branches.

According to expert estimates, in 2006 the turnover from projects employing EPC was around €50 million. The market size in a broader sense, including all "performance oriented" contracts (for instance boiler and heat pump retrofits) was twice as large, around €80-100 million.

Energy saving potential through ESCOs in Sweden has been calculated at 15 percent of the present energy demand, suggesting a €650 million ESCO investment potential with a relatively short payback time.

Clearly, the most attractive emerging sector for ESCOs is the public sector (municipal buildings, hospitals). In less than five years, 5 percent of the public building stock has been contracted by EPC. Two years ago, the EPC growth in this sector had reached about three million m^2 and an additional two million m^2 was under preparation. The most crucial success factor for the uptake of ESCO projects by public bodies is a change in mindsets: ESCO companies have distanced themselves from pure outsourcing, and focused on implementation and operational partnerships instead.

Almost all projects that have been implemented lately have installed new or improved control systems. Large-scale air-handling refurbishment in combination with improved heat recovery accounts for the largest project investment values. An interesting development is that public bodies have also started to use the cost savings from EPC projects to finance RES installations.

Progress and Pitfalls

The Swedish ESCO revival is believed to be the outcome of a complex mix of targeted strategic activities. Key parts of the strategy have been ground studies and market studies, pilot projects and guidelines for procurement and model contracts. In addition, large scale and effective information dissemination and capacity building, combined with personalized information dissemination to EPC buyers has encouraged project development.

As recently as 2005, the financing of EPC was not well established by banks. Today, there is at least one Nordic commercial bank which provides TPF. Smaller projects are directly financed by ESCOs.

Opportunities

The Swedish example demonstrates that deliberate, well-designed dissemination of information, clarified regulatory environment, standardized, trustworthy documents and procedures, and successful show-cases can be of key importance for development. Nevertheless, the recipe is not valid for just any market, different markets have to overcome their own barriers using some of these measures and combined with others.

DENMARK

According to the energy division of the Danish Offshore Industry, fewer than five companies offer ESCO services in Denmark. The number of ESCOs has been rather constant over recent years, although companies enter and leave the market, which results in some small fluctuation. To the knowledge of the author, the ESCO market size has not been evaluated lately, but ESCO experts estimate it to be around €5 million per annum. Experts claim that the commercial market for ESCOs was more favorable in 2006 than it was in 2000. The market is expected to experience a further expansion in the coming years, partially as a result of the governmental commitment to decrease final energy consumption by 1.7 percent/year by 2013. This goal is supported by the obligation on energy producers and distributors to implement energy savings and to document their obtained savings. There is also a new obligation to tender out the energy saving activities thus creating an opportunity for the ESCO.

Until the 2000s, efforts for energy savings and energy efficiency improvements were concentrated on the private sector (industrial sites) and remarkable results were achieved, particularly in the brewery sector. As of 2006, the (public) building sector has been receiving growing attention. The types of projects implemented by ESCOs so far in Denmark have been control system installation, ventilation and industrial process improvements.

According to the Danish Offshore Industry, financing of ESCO projects through banks has not been deployed to date, probably due to a lack of knowledge and experience in the financial sector of the market perspectives of EPC. Clients have financed ESCO projects implemented to date.

Barriers

There is also a need to develop awareness and trust among potential clients through demonstrational projects and making standardized contracts and related documents available. Experts believe that one of the most important barriers to EPC is the lack of established standard monitoring and verification methods. Another major necessity that has been articulated by experts is to establish working networks where utility/grid companies, financial institutions and equipment suppliers can jointly develop organizational and financial models. Today, suppliers of energy saving equipment are often dependent on utilities in order to be able to measure baseline energy consumption and savings.

PORTUGAL

ESCO business activity in Portugal is dominated by seven or eight medium and large ventures and moving upwards slowly. A few of these companies are large multinational ESCOs or daughter companies of the previously monopolistic electricity utility, Energias de Portugal Group (EDP). The number of larger ESCOs and the size of the market have not really changed for several years. New companies do appear, however, and others leave the market or change their core business. There are also small ESCO-like consulting companies that are oriented towards auditing, preparation of plans for rationalization of energy, retrofitting energy efficient equipment, and similar ESCO services.

In spite of the past stagnation of the market, the ESCO concept has recently gained in popularity. The importance of ESCOs is growing as attention is increasingly given to energy savings obligations. ESCO development is also supported by the complete electricity market opening in 2006. It is expected that competition will induce the introduction of more added value services, especially in the case of decentralized energy generation.

Although exact numbers are not available as to the size and potential of the market, ESCOs in Portugal are only targeting a fraction of the market saving potential. Even some typical "low-hanging" ESCO projects, such as municipal street lighting projects, have not yet been fully exploited. It is estimated that 30 percent of municipal energy costs could be saved economically with a short payback time.

ESCO customers are primarily large and medium sized industries and large tertiary buildings (shopping centers, hospitals, hotels). Most attention is given to CHP due to its simplicity, low risk and short payback time, combined with financial incentives (such as high feed-in tariffs) that are given to cogeneration. Activity in relation to renewables has started to emerge over the last few years. Multinational ESCOs also implement heating and cooling solutions as part of facility management. The most popular contractual schemes are the shared savings and chauffage contracts.

The legal framework in Portugal has been supportive of energy efficiency and renewable energies, but not of ESCOs in particular. The CHP sector, which represented 12.2 percent of total national electricity production as of 2003, has benefited the most. In particular, the high feed-in tariffs for cogeneration guaranteed for 15 years have served as an important incentive.

Barriers

In parallel with the positive environment for the development of ESCOs, some significant barriers remain. Financing of ESCO projects through TPF is sometimes in competition with certain governmental support schemes and programs, instead of complementing them.

In spite of the successful examples of TPF and EPC, financial barriers still exist. Return rates are considered insufficient by ESCOs for many potential projects, especially if compared to supply-side investments. Transaction costs are regarded as too high, thus companies still go for projects that they consider more profitable than demand-side intervention. In public building projects accounting rules may override the goal of rationalization of energy use.

As in many other countries, running costs (operational and capital costs) and investment costs are separated in the public sector budgets, and saving on operations does not compensate for the costs incurred in the investment budget. Similar to the situation in Spain, split incentives are also a typical barrier in Portugal.

Tradition and slow uptake of new business solutions have also been reported as a hindrance to the ESCO concept's diffusion. Energy suppliers have long seen themselves as providers of energy *per se* and not of energy services. This situation is, however, changing and large suppliers are starting to offer energy services. Perceived uncertainty of the profits of energy services seems to be another critical obstacle to ESCO investments.

Municipalities do not have enough autonomy to assume multi-year service contracts in energy. The actual law is protecting the already installed utilities to preserve their contracts. The financial institutions do not finance service projects, but are just looking at the companies' balance sheets. Banks are not informed and knowledgeable about the systems and technologies involved and they do not have enough confidence in their market value. This limits the financial resources available to medium-sized ESCO companies. The EPC contract is complex and the law related acts in Portugal are too conservative. Customers' lawyers then doubt the ESCO models and their real application.

Facilitators

Building trust via disseminating information and best practices among potential clients is one of the most important factors that could facilitate the ESCO sector. As of today, the ESCOs active in Portugal deal only with customers, who initiate the ESCO project themselves, while active marketing has been disregarded.

There appears to be a need to integrate the EPC concept into the legal framework. For instance, standard procedures for the planning, implementation and monitoring of a project could be beneficial, documents and guidelines could be developed, demonstration projects provided, and targeted information disseminated by a neutral stakeholder.

Finally, the potential role of the public sector in Portugal is enormous. It is the owner of most service sector buildings, whose energy optimization could serve as an important initiator and multiplier. It would demonstrate the feasibility of ESCOs on a large scale and in front of a large audience. In addition, it would be able to give a basic impulse for the industry, and would set a good example for other building owners.

Government Support

Government actions are creating new ESCO opportunities, including:
1. Since July 2007, a new buildings energy certification regulation, which applies the EU EPDB directive, will make it mandatory for all new and existing buildings to have an energy efficiency certificate if they wish to build it or sell it. This will affect almost 200,000 buildings, which will have to have energy audits and energy retrofits, offering ESCOs an opportunity to apply their services and models.

2. The approved energy efficiency plan explicitly names ESCOs regarding some risk sharing and financial support from the government funds to EPC and also mentions co-investing in the launch of new ESCOs or co-investing in specific ESCO projects with energy effi-

ciency objectives. The plan also suggests the role model that the government buildings should take with demonstration projects. There will be specific funding to support that effort.

3. New regulations now exist regarding a feed-in tariff, which enables the installation of small micro-generation units of solar and micro-wind in homes and small businesses. This promotes the use of decentralized power generation.

SPAIN

This up-to-date look at the ESCO industry in Spain has been provided by Mr. Enrique Gonzalez Roncero, who is with the Union Fenosa utility in Spain. The authors appreciate the special insight he offers here.

In 1974 the "Spanish Centre for Energy Studies" was set up due to the oil crisis and the need to promote energy efficiency and energy savings. In 1984, the Centre was renamed as IDAE, the "Centre for Energy Diversification and Saving," with the function of carrying out energy audits and studies, rural electrification plans, and assessment for energy savings and diversification. The center was also directed to manage funds to subsidize actions eligible for incentives.

At a regional level, some energy agencies were created such as ICAEN (Catalonia), EVE (Basque Country), AVEN (Valencia), INEGA (Galicia) and SODEAN (Andalusia), with the missions of promoting energy efficiency, the introduction of renewable energy sources and investment in projects to develop more efficient technologies.

In recent years, these public agencies have launched a more aggressive promotion of energy services, acting as ESCOs in some cases through creating joint ventures with private companies. The market sectors involved in this promotion were industry and hospitals or medical centers. These projects implemented CHP and HVAC plants, managed plants and sold comfort in buildings. Since 2006 they have intensified these promotions, trying to invigorate the industry and proposing them as one of the main mechanisms to obtain energy objectives and emission reduction.

More recently, private ESCOs began to focus on the industrial sector. Initially, the market was dominated by companies belonging to international groups, which were already providing energy services in other countries. In addition, technology companies began offering energy outsourcing by providing energy fluids and centralizing the energy production.

Over the past three years, the number of companies engaged in energy services has increased and the types of companies has diversified. At present, engineering, maintenance, and manufacturing firms, as well as utilities are sharing and expanding the market. As part of this market evolution, in 2008, the government published the "Plan Nacional de Ahorro y Eficiencia Energética en España 2008-2012," including specific lines promoting ESCOs by activating the Directive 2006/32/EC of 5 April 2006 on "Energy End-Use and Energy Services." The budget for this period is €400 million, in order to boost investment in projects related to energy-intensive industries, large distribution chains, hotel chains, fleets of transport, etc.

In particular, the PNAEE 2008-2012 includes a "Measure on Saving and Energy Efficiency Plans in Public Administrations," according to Directive 2006/32/EC, the public sector must play an exemplary role to improve energy efficiency. This measure must be adopted at the appropriate level, either national, regional or locally.

The energy savings and CO_2 emissions avoided with this directive are estimated at the following values:

	2008	2009	2010	2011	2012	Total
Energy savings (kTep)	14	29	44	59	74	220
CO_2 emissions avoided (ktCO_2)	75	156	236	317	398	1.182

The measures presented in the national plan must be supported by legislative actions with complex administrative changes. In public services, the aim is to reinforce the professionalization and requirements of companies that provides energy services based on energy efficiency (in parallel with quality or environmental stamps to promote the creation of energy efficiency trademarks on their products, processes, services, etc.) in all sectors.

Finally, ESCOs can do good promotion by selling savings and efficiency as well as advocating and engaging their benefits to achieve objectives. It also refers to ESCOs as a type of company that would be a good intermediary for the application of saving policies.

ESCO Industry and Market

Based on the European Commission database, Spain has five companies registered as ESCOs. However, there is evidence that there are additional companies operating as such, and this number is growing.

The ESCO market is constantly evolving to expand the services offered by existing companies, diversifying their services, as well as by increasing the number of companies. However, the absence of an explicit characterization of what should be an ESCO leads to an undefined market. This involves a wide variety of companies that prevent the user from clearly identifying the guaranteed energy savings concept.

This uncertainty, and the fact that most of the companies do not have energy services as a single business line, results in a lack of data on the energy services business in Spain.

Types of Projects and Contracts

In Spain there is no database of EPC projects; therefore, the information available gives a qualitative vision of the main lines of business. In this regard, the government has been promoting energy services in public buildings, industry and the tertiary sector. This has marked the type of projects that have been developed.

The ESCOs in Spain offer, generally, three types of projects covered as energy services:

- Expanding energy audits, offering implementation and financing of some of the energy conservation measured discussed in these studies. In some cases, also guaranteeing savings.

- Expanding the provision of maintenance of thermal machines, offering guaranteed savings, providing preventive maintenance support and offering proper management of energy. In some cases, it includes investment in new equipment.

- Implementation of fluid supply systems, maintenance of facilities and energy management (district heating-cooling, cogeneration).

At the moment, shared savings and guaranteed savings contracts have not had much penetration into the Spanish market despite the upward trend of energy prices. Recently IDAE began to develop an "energy services" contract to be used as a model.

The type of contract mostly used in Spain today is the "energy supply" contract. Through this method, ESCOs have launched large photovoltaic installations, with areas larger than 1,000 m^2, keeping the ownership and just selling energy produced. At the end of the contract, usually between 12 and 15 years, the client becomes the owner of the solar thermal power plant.

The client can finance the plant by himself, by external financing or subsidies that encourage the use of renewable energy. In this scheme of financing, these contractual terms acquire special relevance:

- To include a minimum consumption by the client, usually between 80 to 85 percent of production capacity;

- The price of energy supplied, being lower than the starting price, includes;
  — The depreciation of the facility;
  — Maintenance;
  — Cost of supplies to support (water, electricity);
  — The profit; and

- The monthly billing is done by a fixed amount, settled at the end of the year according to actual consumption.

Barriers to ESCOs and EPC

Being a new service for the Spanish market, there are quite a few barriers that must be overcome and can be classified into the following:

- Legal:
  — Adapted from EPC to Spanish law;
  — Adapted to the special conditions of each client;

- Economic and Financial:
  — Lack of capital and financial mechanisms for such contracts;
  — Difficulties in obtaining money from a financial institution for lack of successful examples in the Spanish market;
  — Projects economically unattractive;
    - The price of energy is not high enough;
    - And the price of equipment is high;

- Commercial:
  — To publicize the service to potential customers;

- Customers resist paying higher margins than they are used to;
- Difficulties in developing a project that is favourable for both the client and the ESCO;

• Directives:
  - Lack of managers' motivation;
  - Lack of awareness on energy efficiency opportunities and lack of experience in efficient technologies;
  - Energy is considered a fixed cost;
  - Cost of energy is irrelevant compared to other fixed costs of businesses; and
  - No trust in savings and incomes proposed.

Future Expectations and Factors Affecting ESCO Development

People are now more interested in energy efficiency projects and are starting to understand the ESCO concept. It is also hoped that regulation and policy support of the government will help to develop the ESCO industry.

The trend of increasing energy prices and the "free market" for selling electricity and natural gas (the "regulated market" has just partially disappeared) will help to implement energy efficiency projects.

Case Studies

The following case studies are in their initial phases, but illustrate the potential in Spain.

Client Outline of Projects

Hotel located in Galicia, Northwestern Spain	**Client:** Hotel belonging to one of the 10 leading hotel chains in Spain.
	ESCO: Unión Fenosa.
	Measures: Optimization of lighting system, optimizing the air conditioning system, carrying out improvements in indoor and outdoor pools (air conditioning and insulation).
	Contract: Has completed the study. Still to implement cost-saving measures identified.
	Project Costs: €205,000.

(Continued)

Client	Outline of Projects
	Annual Saving (energy and water): €50,000.
	Emission Reduction: 284 CO_2 tons per year.
Hotel located in Canarias Islands	**Client:** Hotel owned by one of the top 5 hotel chains in Spain.
	ESCO: Unión Fenosa.
	Measures: Installation of a new chiller with heat recovery, optimization of the lighting system, installation of variable frequency extractor kitchen, heated pool programming schedule.
	Contract: The study is being conducted; energy detail is needed for a definitive determination of the definitive measures and savings.
	Project Cost: €227,000.
	Annual Saving (energy and water): €40,000.
	Emission Reduction: 95 tons CO_2 per year.

Chapter 3
Eastern Europe

There is always a danger in losing the uniqueness of a given country by regionally grouping it with other countries. This danger is even more pronounced in Eastern Europe where the countries range in size from Slovenia, which cover 7,820 square miles to Russia, which spans 10 time zones. But these countries do have much in common as they are steeped in old customs and traditions, marked with rapid political change and have exceptional potential economic growth. A history of planned economies typically means there is an absence of data to establish baselines. Government subsidized energy prices have been common. Old building stock typically suffers from deferred maintenance and needs renovation.

A cursory look at the respective economies in Eastern Europe reveals trade liberalization, investments in infrastructure, and other indicators which typically enhance economic growth.

One multi-national company, Dow Chemical, in doing a market analysis, grouped 31 countries into what it calls the East Europe Growth Region (EEGR). While the 500 million people in this area have a wide range of cultures, religion, income levels and political traditions, the company stresses its promising future and how quickly much of it is transitioning from "developing" economies to "developed" status. In 2007, the region's sales exceeded $2.6 billion, a 30 percent increase from the prior year. The 2007 growth in Russia and the CIS over the 2006 totals was an even more impressive 50 percent.

In an area where certain commonalities are obvious, there are clearly many contrasts. In the history of energy service company (ESCO) development there are countries where the ESCO industry emerged in a very short period at the onset of the 21st century. Austria and the Czech Republic became the new success stories by 2005. On the other hand, there were also some negative examples, where EPC failed and potential ESCO development was thrust back even farther due to a lack of trust. This happened in Slovakia and Estonia. Finally, a group of countries could be characterized by low level ESCO activity in the early years of the 21st century due to internal and external factors that had prevented prior development. In the case of Lithuania, energy efficiency has been a priority, but marketing channels other than ESCOs have delivered it. On the other hand Greece, Poland, Romania, and Bulgaria have been examples where large potential for energy savings exist, but still little energy efficiency activity has been undertaken until recently.

Portions of this chapter have been revised and updated from the 2007 EC ESCO report previously cited. Where contributing authors have offered us special insights, they are identified at the beginning of the narrative for the designated country.

CZECH REPUBLIC

The Czech Republic is the ESCO frontrunner among the Eastern European countries, even though the market is still considered to be in its initial period. The EPC concept was unknown in the Czech Republic until 1992, yet in just three years the country saw a rapid takeoff of ESCO activity with a €3 million investment in energy efficiency in the public healthcare sector.

The first project was the renovation of the thermal energy handling system at Balovca Hospital. In fact this was the first EPC in any Eastern European country and was brokered by Kiona International (later part of Hansen Associates) and conducted by an American ESCO, EPS, in partnership with Landis & Gyr (now Siemens Building Technologies). A key support leading to the project's success was the participation of the energy center, SEVEn, and Ivo Slovotinek in particular. The project saved so much energy the first year (43 percent) that the work was renegotiated to create a more comprehensive project. In an unprecedented move, the facility director, who lead the hospital's ESCO efforts, became head of the hospital.

Until 2001, ESCO development was slow due to numerous barriers and obstacles, but in that year the Czech market reached a turning point because of important changes in legal circumstances. A new law was passed which made energy audits obligatory for large consumers. This decision has meant a strong push for energy efficiency investments. In 2004 the State Energy Policy was adopted, which highlights the role of energy efficiency. Accordingly the National Programme for Energy Effective Management was accepted, where EPC is recognized as one of the support mechanisms for energy saving.

According to the Czech ESCO database, managed by the Czech Energy Agency, there are currently five companies in the Czech Republic that are focused on providing services according to the EPC concept, and another two companies work as ESPs. However, the SEVEn database reports 10 to 15 ESCOs in operation. This number is increasing. In 2005 alone two new companies were created for the provision of ESCO services. There are also approximately a dozen companies providing long-term energy delivery contracts.

The potential of energy savings through EPC, which are economically attractive, is about €100 million. Available estimations of the market potential vary slightly, but they are in the range of €10-20 million/year. Until now about 70 projects have been realized through EPC but over 30 percent of these have been done by one ESCO. The most effective tool to promote ESCOs in the Czech Republic has been large scale programs to raise awareness, where the ESCOs' own activity has been particularly valuable.

ESCOs' successes across sectors are varied. The healthcare sector is the primary focus for ESCOs, while educational buildings, military and other state owned sectors are appealing projects, too. Military refurbishment projects are complicated with special legal conditions, but the interest is high due to the high energy saving potential.

Middle-sized cities are very active in working with ESCOs, bundling buildings into project pools. Apart from the public sector, the private sector (typically industry) is also on track regarding energy efficiency investments. The rough division between ESCO investment categories is shown in Figure 3-1.

The most common contracting form is guaranteed savings. The market is diversifying and consolidating. Contracts are individualized as a result of negotiations between ESCOs and their clients. Financial institutions, including mostly local banks, are available and are ready to participate in TPF, but multinational ESCOs often use their own corporate financing. The International Financial Corporation is running its "Commercialization Energy Efficiency Facility," providing loan guarantees for ESCOs and end-users. There is also limited governmental support for ESCOs thorough the Czech Energy Agency.

Barriers

Despite some consolidation and an overall increase of the market in recent years, some barriers still exist. Skepticism by management towards

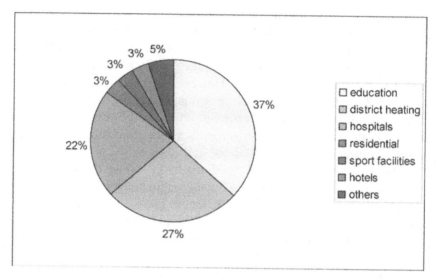

Figure 3-1. Distribution of EPC in Investment Categories, 1995-2005 (GreenMax-Capital Advisors for IFC, 2006 from Dasek pers.com.)

energy efficiency investments and EPC still lingers as a result of earlier unclear definitions, confusing concepts and some failed contracts. Correct understanding of the benefits and conditions of ESCOs and EPC has grown, but prejudices against complex solutions remain strong. Increasing effective information dissemination, raising awareness, and education related to energy efficiency is still essential to overcome this problem.

On the ESCOs side, companies are not yet ready to take projects with a long payback period. Typical projects are four to six years long, and the majority of ESCO investment interest is for heating equipment (heat delivery regulation, piping, pipe insulation, boiler replacement, fuel switching), or power factor management. However, most industrial end-users have already installed such equipment. On the other hand, insulation and other building renovation measures have a long-term pay-back period and are therefore, excluded from the current scope of attention. It is considered too risky to invest in long contracts in the private sector because of the unpredictable financial future of potential clients. To a lesser extent, the same fear is present towards the public institutions. Governmental support for EPC in the public sector would be necessary through the provision of guarantees of long-term standing to potential clients. Nevertheless, longer projects are already starting to gain importance in the public sector, especially for street lighting and energy delivery.

Detailed, reliable information on present energy consumption and condition of buildings is lacking, which hinders the easy set up of energy efficiency targets. It would appear that more attention needs to be given to appropriate project implementation and especially monitoring and verification of savings.

Legislative barriers have largely been removed both in the public and private sector. Standardization of public tenders and verification of EPC contract evaluation procedures is vital. The responses to calls for tenders often include both EPC and EC. However, public administration is usually unable to effectively compare these two different types of proposals, which causes confusion. Standardization of EPC project procedure would be particularly beneficial to avoid dissatisfaction, and unwillingness to start the process and to help ensure completion of successful projects.

The country can already present success stories and areas of well-developed energy efficiency focus. However, there is still room to develop, and some sectors with high saving potential have hardly been tapped (such as the military). The legislative background has become exemplary (obligatory audit) and has advanced the sector significantly, although some important issues, such as procurement, are still open for change.

POLAND

The authors are indebted to Mr. Janusz Mazur, president of the Board of the Energy Saving Company ESCO Ltd., of Krakow, Poland, for a very insightful and candid analysis of the current ESCO situation in Poland. Mr. Mazur identifies several factors holding back ESCO development and some potential remedies.

ESCO Development in Poland

The documented history of ESCO projects in Poland started at the beginning of the 1990s. It was then that, together with democratic changes, the possibility of making contacts with institutions and companies from countries, which were already aware of this model, occurred. The very first projects were small and rather demonstrative. Among the pioneers of ESCO projects were MPEC (Municipal District Heating Company in Krakow) and the Ministry of Defense. MPEC in Krakow was then active in two international projects. The first was conducted by the World Bank—Modernization and Restructuring of the Heating System. The second was implemented together with the U.S. Department of Energy's Krakow Clean Fossil Fuels and Energy Efficiency Programme.

It was then that the first company was established (ECOGY Ltd.), with a goal to conduct a project for the modernization of coal boiler houses according to the ESCO model. Its co-founders were JAIDO—Japan International Development Organisation, Tecogen—an American company, MPEC Krakow and Naftokrak-Naftobudowa—a building and construction company from Krakow. ECOGY conducted the modernization of a coal boiler house in Wieliczka near Krakow, which was designed to be re-paid by the savings achieved by the client.

In the late 1990s, the ESCO model became a rather commonly used method for the modernisation of heating systems at military units. The Ministry of Defense made a huge effort to organize the financing for this modernization effort; and, according to this model, included an adjustment of the strict financing rules for budget units with military secrets. At the time, this was the only major project that was implemented in Poland on state property.

About this time, large global and European companies began to open their branches in Poland with the intention of running the business according to the ESCO model. At least the following companies should be mentioned here: Dalkia, MVV, Landys & Gyr (later a part of Siemens), Ineoineo, and Auxima Services.

Dalkia's business beginnings were rather important because this company took an unusual systematic approach to the ESCO project organization. Dalkia was interested in the Polish market by a private installation company, ZIS Ltd. from Krakow (today ZISTECHNIKA Ltd.). It also became the co-founder and shareholder of the Dalkia Poland company. The second co-founder and shareholder, besides Dalkia, was the European Bank of Reconstruction and Development. The main range of modernization activity was, at that time, small and medium heating systems.

ESCO projects at the turn of the century were also conducted by companies that did not have performance contracting as the main scope of their activities, including the DEng Ryszard Śnieżyk's engineering company from Wroclaw and the Prowinex company from Krakow. Other companies that were founded around that time were: Harpen, Bioenergia, and Finesco, which was founded by the Polskie Sieci Elektroenergetyczne (Polish Electro Energy Network).

MPEC in Krakow, mentioned earlier, brought into life in April 2000, the Energy Saving Company ESCO Ltd. (POE ESCO). The company was founded as a result of World Bank inspiration, which had been financing Krakow heating system modernization since 1992. POE ESCO was created as a company for the complex thermal modernization of buildings and the modernization of heating systems as a specialization. The World Bank took a huge part in the company's development, indirectly financing studies, training and research, and activating GEF funds for investment activity and technical support.

ESCO Market in Poland

Today the ESCO market in Poland is rather small. As a result, not many companies conduct performance-based contracts. Today, the number of companies that have active performance contracts is estimated to be between 8 and 13. Among equipment suppliers, the most active is Siemens. Honeywell, unfortunately, has recently limited its activity as far as ESCO projects are concerned.

Among the municipal companies, the most active are MPEC Krakow, which conducts projects via its own company POE ESCO, and ECO Opole (Energetyka Cieplna Opolszczyzny—Opole Province Heating Industry—a heating company from Opole) which also has its own projects. French Dalkia and German MVV are active in municipal projects. Today, Dalkia Poland is a part of the Polish energy and heating industry on a large scale and pays less attention to ESCO projects for small heating systems. MVV focuses its activities on the modernization of productive systems by assembling small gas-powered CHP systems. Other companies that have successfully implemented ESCO projects are Harpen and Bioenergia. Among engineering companies, the most active is DEng Ryszard Śnieżyk's from Wroclaw.

Contracts and Financing

The main business core in Poland for ESCOs is building and construction work, which should lead to their customer's lower energy consumption. It is rather rare to contract power supply. The main cause is the complexity of the Energy Law. Real estate companies often calculate and sell the energy indirectly. Both power supply contracting and "chauffage" is possible in Poland in a limited manner.

Most contracts with a savings warranty in Poland have probably been signed by POE ESCO from Krakow. So far the ESCO has 8 contracts including 32 buildings by different owners. Warranties concern the thermal energy consumption used for heating, based on the historical base value as a three-year average. This procedure is also obligatory for such customers as the Polish Armed Forces.

The main financing sources have been commercial banks, private sources, financial institutions and, until the end of year 2008, the World Bank. For thermal modernization, while the repayment period in Poland oscillates between three to five years, it is necessary for the customer to make a rather significant contribution (the first large instalment) in the amount of 50 or even up to 80 percent of the entire project value.

Project Types and the Market

ESCO projects that are conducted in Poland actually include all of the market sectors. Among the property of the local authority, the most often implemented performance contracting undertakings in the ESCO model are the thermal modernization of buildings (insulation of walls and roofs—thermal isolation, window replacement, modernization of the central heating system), replacement of heating sources and the modernization of street lightning (most often the replacement of mercury lamps for sodium).

The Ministry of Defense put ESCO projects for modernization of boiler houses at military units out to tender by a company, "Energia," that administers the power supply for the Polish Armed Forces.

Undertakings implemented in the industry sector have a diverse range: from the use of waste energy to efficiency improvement for energy generation and thermal modernization. Increasingly, ESCO companies undertake, in the industry sector, projects including the building of modern, small, gas-powered CHP (currently the main activity range for MVV and Siemens). POE ESCO has recently undertaken a project for local transport replacement of oil-fuelled vehicles to those driven by compressed gas.

The largest thermal project based upon the ESCO model was put out to tender by the city of Łódź (approximately 400 buildings). However, it has never been settled. The army has gradually modernized the heating systems (six to eight systems a year) at military units and the contractors must provide documentary evidence for the results of the respective savings level. A huge project (28 educational buildings) was conducted by the municipality of Krakow.

Industry has had real difficulty accepting the ESCO model. Often the bills for energy or lighting are only a fraction of total production costs. Members of the boards are rarely interested in fractional percentage savings as they follow a rather strict formula. This situation has also been negatively intensified by the attitudes of those responsible for the energy management at the production plants, who are not willing to admit their mistakes at work. There are, however, some positive examples in this field; e.g., the long-term co-operation between DEng R. Śnieżyk and Polar (household goods producer).

In the commerce sector, the bulk of the work has been conducted by the TESCO network on its own as far as the optimization of electric power contracting is concerned.

Some interest in the ESCO system has been shown by the health department. After implemented reforms, they have become an example of budget financed units that are able to retain all the savings obtained. However, successful undertakings are, today, still rare.

Limitations for ESCOs and EPC

The Polish legal system does not prohibit ESCO activity. However, it lacks references or clues concerning the organization or financial settlement for this kind of business activity. It leads to enormous reluctance and apprehension among the potential customers regarding the use of this procedure to finance and implement projects. The government does nothing in this case. It is difficult to expect that even five active companies will achieve an effective market for this kind of undertaking in a country with 40 million people.

An analysis of the performance contracting market deterrents in Poland has several dimensions; 1) the ignorance and misunderstanding of this model by potential customers; and 2) a lack of outsourcing traditions. The ESCO company is after all the general contractor for several services: engineering, construction, and financial services; and 3) the lack of understanding of the risk that ESCO companies take on, and its value for the clients.

The work of R. Śnieżyk, Modernizacja Ogrzewania Szpitalnego metodą ESCO (Modernization of Hospital Heating System with ESCO method), indicates ESCO companies in Poland are exposed to a number of risks that can vary from those in other countries, including:
- strict activity of the Inland Revenue;
- changes in the legal regulations (especially tax, energy, and environmental laws);
- "sabotage" from the customer's employees who are afraid of being fired;
- customer's bankruptcy, which is protected in Poland;

- last place on the customer's list of creditors—meaning potential problems with debt collecting; and
- difficulties in estimating a share in the ESCO undertakings in the results that the customer gains also from other directions.

Other limitations arise between tax law and accountancy regulations. The first problem is the necessity of paying for taxes immediately after issuing the invoice, regardless of the re-payment period for the client. For the combined sale of construction and financial services (on the invoice the total financial cost value for the whole period is counted for the construction works) the ESCO must pay enormous sums in VAT. In practice, this increases the cost of the entire service as a whole and lowers ESCO company's liquidity.

Another major problem is also caused by not correctly defining the term of ownership for the improvements and equipment that have been accomplished by the ESCO. Usually they are owned by the ESCO until the whole amount has been paid. However, because the company has already invoiced (sold) them, it cannot claim capital allowances for those fixed assets. The inscription in the agreement (a kind of security measure) which states that the equipment is owned by the ESCO until totally repaid, makes the case very complicated. Who is then authorised to claim capital allowances?

An important area that restricts the efficient market has been the lack of functional financial instruments that are specialised in ESCO companies. Until recently even those easiest and most popular; i.e., guarantees, were practically out of reach. The companies that develop most actively in Poland are those with significant ownership capital (Dalkia, Siemens, MVV), or those with an organised guarantee system such as POE ESCO (the guarantor for all the lines of credit for this company is its parent company MPEC SA in Krakow—a huge local heat supplier). The lack of an easily available guarantee system caused VERST Energy Consulting from Bielsko Biala to withdraw from the ESCO market.

There are also programmes that compete directly with energy performance contracting. The largest of these is the national subsidy programme for thermal modernization, as determined by an enacted bill. The easy method for gaining a donation for thermal modernization competes efficiently with ESCO companies' offers. The same competition applies for subsidies from the European Union programmes. The ESCO market is also weakened by unfavourable relations between the prices for construction work and energy costs, which are in practice regulated by the government.

Therefore, most limitations are the result of either the complete lack of interest from the government of Poland for these services, or the provisions being excessively liberal. It is clear, unfortunately, that this has not provided the desired results. The tertiary sector of the economy for ESCO services is nearly nonexistent.

Chances for the Polish ESCO Market

In a country with so many limitations for the development of the ESCO market, there is a wide range of possibilities to improve this disadvantaged situation. First of all, Directive 2006/32/EC of the European Parliament and of the Council on Energy End-use Efficiency and Energy Services, should bring positive results in the future. The directive defines ESCO-type activity and assigns an obligation for each country to ensure that there is enough encouragement, fair competition, and equal operating terms for ESCO entrepreneurs. Moreover, it would help if the state were to create special funds, which could be made available to ESCOs for subsidies, loans, financial guarantees, and other financial measures that are related.

The directive forces upon all member states the obligation of preparing for an energy efficiency action plan (EEAP). In the document drawn up by the Polish Ministry of Economy, ESCO companies are mentioned several times. The document is based upon the assumption that the ESCO model is going to be one of the measures used to improve energy efficiency. It even takes into consideration promotion of the energy services provided by ESCOs from 2009 to 2016 to stimulate the market. One should hope that those documents will lead to some activity from the Polish government in this field in the near future.

The next positive contribution for improving the unfavourable situation of ESCO companies in Poland is the reorganization of the Partial Loan Guarantee Programme with Bank Gospodarstwa Krajowego (National Economy Bank). The programme was created in 2005 from GEF (Global Environmental Facility) funds financed by the World Bank. Unfortunately it has not, in its first edition, awakened much interest from the market, companies or banks. After some changes made recently, there is a chance now that it can become a very useful instrument.

Other easy possibilities which could stimulate the ESCO market; e.g., include the possibility to gain more points for including the ESCO model during the evaluation process for investment projects that are financed by European Union programmes. This is the example of an activity that does not cost a penny.

It would be nice, as in other countries, to introduce some tax or capital allowance or similar measure while implementing ESCO undertakings. All of the ideas can be multiplied endlessly. Undoubtedly, it would be interesting to create an independent entity—a facilitator that could help in putting the ESCO energy efficiency project to bidding. The choice of method and strategies belongs, of course, to the government, which unfortunately is not paying much attention to this solution.

CASE STUDIES

Szaserow Hospital	**Client:** Szaserow Hospital, WIM (Military Medical Institute), Warsaw (2005)
	ESCO: Siemens Ltd.
	Range: modernization and automation of a coal boiler plant (steam) with auxiliary equipment
	Contract: 5 year savings warranty
	Type of contract: Public, 5 years
	Project costs: €630,000
	Annual savings: fuel and electricity €130,000
	Warranty for boilers efficiency 78% (efficiency before modernisation—55%)
	Emission reduction: CO_2: 3762 t/a
Thermal modernization of school buildings	**Client:** Kraków Municipality (2004-2007)
	ESCO: Energy Saving Company ESCO Ltd.
	Range: modernization of DH systems (radiators, thermostatic valves), window replacement, thermal insulation of roofs and walls, and (partially) equipment
	Contract: 10 year savings warranty,
	Type of contract: Public, 10 years, EPC
	Project costs: ca. $11.9 million
	Annual savings: heat 6.8 million MWh—ca. $610,000, payback period ca. 20 years
	Emission reduction: CO_2: 373t/a

ROMANIA

The Romanian ESCO market is in an embryonic state, with few companies willing to become active. The Romanian Energy Efficiency Law was passed in 2000, which puts forward a number of measures to support energy efficiency. International agencies (EBRD, USAID, World Bank/GEF and UNDP) have also been active in the development of energy efficiency financing procedures. Training in EPC was funded by USAID as early as 1992. Energy Efficiency Fund 63, was established by the World Bank and financed by the GEF and the Romanian Government to increase energy conservation activities and implement measures in the country. In spite of these efforts, the ESCO market has not been able to get off the ground because of a number of strong barriers.

Currently there are two companies—one specializing in electricity and the other in thermal services which qualify as private ESCOs and offer pure EPC solutions. One of them was set up in 1996, thus becoming the first Romanian Energy Service Company, and it has 100 percent local Romanian ownership. The other ESCO was founded in 2004 as a Romanian-Canadian Joint Venture. These companies offer a wide range of services, including ESCO projects, auditing, equipment installation, operation and further engineering projects. In addition, there is one ESCO-type company chiefly working with CHP projects. There are a few regional ESCOs active in Romania offering energy supply contracting. ESCOs also do boiler renovation and the operation of residential district heating.

The UNDP/GEF Energy Efficiency Project was launched in 2005 as a collaboration of four partners: UNDP, GEF, Agentia Romana Pentru Conservarea Energiei, and the UN Office for Project Services (executive agency). The program assists energy efficiency investment in various ways. First of all, technical assistance is offered to the public and private sector through feasibility studies. For the public sector a direct contribution to equipment purchase is also available of up to 20 percent of the investment but is limited to €50,000. Finally, the so called "deal building" brings together energy efficiency investors and financiers and offers advice when needed. These support activities have been able to catalyze large-value energy efficiency projects—over $20 million investment so far and about $7 million in the pipeline. Besides international financial institutions, local banks have started to move into the energy efficiency market; however, they have not yet fully acknowledged the potential and they are not completely familiar with the ESCO concept.

Banks still lack the internal expertise to evaluate energy efficiency projects. They are not yet ready to finance projects based on the credibility of the investment and the prospects of the savings, but still use the traditional asset based financing, and evaluate the client's creditworthiness.

In the municipal sector the lack of off-balance sheet solutions is the major obstacle. The municipality cannot take the investment of the energy efficiency equipment on its balance sheet because it would often override the maximum lease credit (given in percentage of total budget). The ESCO investment is not a traditional loan, so it should be treated differently. The ESCO is also unable to claim the investment on its own balance sheet, because after a few projects no bank would lend to it.

Industry is also a likely client of ESCOs and in fact some EPC is implemented there. Audits are obligatory above a certain size, and this should make the EPC market more active. However, penalties are so small that plants prefer to pay them than comply with this regulation. Industries are also reluctant to give out data about their sites.

After reviewing the above material revised from the 2007 ESCO report, Mr. Florin Pop, of Energobit, added the following observation regarding the ESCO industry in Romania.

> Due to the fact that in the Romanian legislation there is not any definition of an energy service company (ESCO), it is not very clear to the customer what exactly an ESCO can do and what are the possibilities of an ESCO/EPC contract. This is the reason why a lot of companies that provide EPC solutions call themselves ESCOs.
>
> The UNDP/GEF Energy Efficiency Project was a success, but now the project has been finalized; ARCE (the Romanian Energy Efficiency Agency) is a permanent contract with Berliner Energie Agentur from Berlin and tries to adopt the model of an EPC contract, developed by Berliner Energie Agentur, to the Romanian legislation and at the same time provide this model for companies interested in EPC contracts.
>
> Hungarian ESCOs are interested in the Romanian market and they are active especially in the Transylvania area (near the Hungarian border).
>
> The legislation in the public sector has not been modified, so there are still difficulties in realizing EPC contracts in that sector.
>
> A lot of private companies are interested in realizing an energy audit. As a consequence of the Energy Efficiency Law and the fact that managers of

these companies understand the need and the importance for realizing an energy audit.

There are not any technical barriers to realize EPC projects in Romania as all the new technologies are available; the financial barriers, however, remain.

HUNGARY

The development of the Hungarian ESCO industry has been celebrated as a quite unique success story not only in Central Europe, but also across the EU. The ESCO industry in Hungary dates back to the early 90s when Hansen Associates provided EPC training funded by USAID.

Based on a registry of the Energy Center, there are about 30 ESCOs or ESCO-type companies in Hungary, but five or six companies cover 80 percent of the market. French ESCOs played a crucial role in the early development of the industry in Hungary; another important factor was strong local engineering expertise and interest in entering a new market for energy services. The ESCO markets size is approximately €150-200 million, excluding large power plant investment opportunities.

Projects in the beginning were primarily focused on public lighting, cogeneration and district heating system improvements. Recently, other technologies have been gaining increasing importance, such as heating and hot water system interventions, industrial water and steam supply, air conditioning, automation and biomass.

Most of the clients have been in the municipal sector. This is partially due to the long-term security that this sector provides for ESCOs, and to the specific support programs. These include the UNDP/GEF Hungary, Public Sector Energy Efficiency Project, and the Szemünk Fénye (Light of Our Eyes) program. The Phare cofinance twinning projects, the German Carbon-Aid Fund, which targets energy efficiency in this sector, have increased the willingness to employ ESCOs. The involvement of ESCOs in the residential sector is also possible, although only through the combination with targeted state subsidies and/or subsidized loans.

After 15 years the market is experiencing some stagnation. The easy "cherry-picking" projects have already been exploited and the market is in need of revitalization. The ESCO sector is undergoing a transformation process with some companies exiting or changing their core business away from energy service provisions.

Strong barriers still restrict the expansion of residential projects. The industrial sector has been gaining more attention lately. Recent figures suggest the distribution of ESCO projects is 30 percent in industry, 30 percent in district heating retrofits and development, and 30 percent in the municipal sector. Renewable energy investments have been started, although these have not gained a major role yet. Projects had typical payback times of three to five years during the 1990s. Today this figure is five to seven years, which is actually one of the major challenges the ESCOs are facing. Companies engaging in projects with longer payback times have to be financially stronger and more stable than previously.

In parallel to the increasing time frame of investments, international aid, which was previously very substantial in Hungary, is decreasing or coming to an end (for instance UNDP/GEF Hungary Public Sector Energy Efficiency Project). This, in principle, should not be a problem since they were intended to aid the development and establishment of the sector.

Barriers

Experts have repeatedly highlighted the problems caused by the lack of baseline data. Sites and buildings sometimes do not have detailed billing systems and pay average fees per month, not according to the real consumption. Thus, ESCOs are in a situation where they cannot prove the savings achieved using this information, or they have to spend one to two years before the project begins, establishing the baseline information prior to project implementation. In this situation, the energy saving potential (and whether an EPC project is feasible) could be evaluated only after one to two years already had been spent on the case. ESCOs must use different ways to estimate the savings. In addition to this, accepted M&V practices have not been widely introduced. Trust established between an ESCO and the client is needed for the remuneration of an ESCO project.

Availability of financing can cause problems both in the municipal and residential sectors. Municipal borrowing is restricted by a cap on obligations creating debt. Although, an ESCO project is not a traditional loan because the savings appear every year, in legislative terms there is no difference. It would be especially beneficial to impose different accounting rules for ESCO projects taking into consideration the repayment of the investment.

Similar to many other countries, most local authorities are still not informed about the opportunities ESCOs offer and are often suspicious of the financial schemes. In addition municipal authorities often feel uncomfortable about sharing the financial benefit of their project with a private company and, as a result, a project gets postponed or never implemented. Procurement difficulties (only cost-related criteria), and fear of having redundancies, also limit municipalities' willingness to engage in ESCO projects. With the four year election cycle, it is particularly difficult to plan in the long-term and to conclude ESCO contracts that are longer than three to four years. Finally, split incentives with investment and operational expenses paid by different budget lines are still very significant, and have always been a major obstacle.

Decision making and the conclusion of a long-term contract is very hard in the case of a large block house, where the law requires the consensus of all apartment owners. Some projects (façade renovations) may be done with the agreement of only the concerned apartment owners, or renovations can be carried out only on the apartments which agree to it (and finance it). This is not possible with, for instance, hot water, heating, or insulation renovations. Furthermore, ownership of certain objects (the water tubes, walls) is not clear (not stipulated in the housing association contract) and can cause a stalemate.

Some of the above barriers are found in the industrial sector, too. The lack of baseline data and the difficulty of defining the scope of the projects because of the complex structure of the plant systems is the most important.

Revision of problematic legislation (ownership-related issues) is therefore desirable, while proper enforcement of other existing laws (notably the obligation for renovation funds in the residential sector) could also help to bring stronger involvement of ESCOs in the residential sector.

Opportunities

Many positive examples have been seen, where personal commitment of energy managers at municipalities has been a significant catalyst for ESCO projects. Also, the importance of the UNDP/GEF Hungary Public Sector Energy Efficiency Project, which has played an important role in the ESCO project development in this sector has been significant.

The residential sector could play a much bigger role in practically all CEE countries, especially with rational utilization of new state support programs.

Dissemination of information remains important. Although there is a growing understanding in the way ESCOs work and the benefits they deliver, the level of acceptance and trust is still considered as one of the major (if not the most important) impediments. Well-disseminated demonstration projects, establishment of an ESCO association, and finding ways of explaining the short and long term benefits to decision makers would serve the ESCO industry well.

Financing of ESCO projects by banks is not a problem. On the one hand, some (mainly multinational) ESCOs have sufficient financial means. On the other hand, third-party financing is a well accepted and widespread scheme. Banks are particularly open to participation in performance contracting.

It is widely believed that actions should be taken to overcome the long-standing barriers to ESCO projects by finding ways to support the market itself, not necessarily individual companies or groups of companies. There is growing support for governmental action based on a combination of appropriate legislation, regulation, monitoring and enforcement, and that these actions be combined with extensive and innovative information campaigns.

SLOVAKIA

ESCO-type companies appeared early in Slovakia. Even though energy service companies began in the early 1990s, and struggled in the beginning, 2003 could be considered as the real starting point. By 2006, there were about 30 ESCOs and ESCO-type companies.

The Energy Center in Bratislava has divided companies providing EPC into four categories according to their orientation and potential to offer ESCO services. These are:

1. Foreign companies offering a full range of solutions, including ESCO equity financing, who are the most active and successful in the traditionally defined ESCO business.

2. Foreign based companies whose main profile is not energy efficiency service provision, but energy systems operation, such as district heating operators. These are not genuine ESCOs and most of them are public-private-partnerships (PPP), generally in the form of joint ventures of private companies with municipalities.

3. Local ESCOs in Slovakia have their own capital, expertise and know-how, but are focused mainly on small energy systems where the costs

of reconstruction are feasible with a limited budget. In many cases they apply the BOOT scheme.

4. Potential ESCOs, energy system operators, equipment and engineering companies, which are not yet able to offer financial services due to their small equity size and the lack of good financing through banks.

ESCO clientele has, until now, included municipal buildings, schools, banks, and hospitals; while outsourcing in the industry, and private facilities is becoming more and more popular. ESCOs normally participate in building renovation and public lighting projects.

Barriers

Similar to other countries with formerly planned economies, the lack of data to construct baselines, subsidized energy prices, and poor management/operation of buildings are common. At the same time, 70 percent of the building stock is in need of renovation. Furthermore, ESCO contracting in the municipal sector is hard to carry out administratively, because public spending requires tendering and comparison of at least three offers. Often the number of companies able and willing to participate in the bidding may be lower. Therefore, energy saving potentials are extremely high, but the need for certain legal, institutional and social adjustments is obvious.

The Slovak banking sector still has limited understanding of, and experience with, EE project financing. Banks therefore, perceive such projects as risky, resulting in lending terms that may not be acceptable for ESCO project developers. This limits ESCO activity to the large ESCOs that can financially support their own activity, and could limit the growth of the sector in the medium term.

Opportunities

ESCOs' opportunities are expected to grow, in line with the general energy conservation and energy efficiency requirements as governmental commitments have been emphasized in the 2005 National Energy Policy. This policy identified many tools promoting EE, such as minimal requirements for energy efficiency of new and large existing buildings. The required introduction of regular controls of the boilers and air conditioning systems of certain buildings, and the introduction of energy certification for buildings has acted as an incentive.

CROATIA

The estimated potential for energy savings in Croatia exceeds €400 million. There is currently only one established energy service company in Croatia offering EPC, which was established in 2003, and is a recipient of financial support from GEF and, indirectly, World Bank loans. There are new companies that are attempting to enter the market, and there are many energy efficiency provider companies that do not offer a guarantee. The total amount of ESCO investment has been increasing in the last few years.

So far three projects have been completed in the country; these have been focused on public lighting and system improvements in educational buildings. In the most recent year there were more than 40 projects in the preparation or implementation phase. These projects focused on a number of technologies: public lighting, cogeneration, HVAC, steam-system recovery, and insulation.

The sources of financing varies. Besides international aid and loans (World Bank, GEF), local financial institutions have proved to be interested, and the ESCO's own equity is being used for project implementation. There are further funds and programs in Croatia for energy efficiency, such as the Fund for Environmental Protection and Energy Efficiency (in the form of subsidies) and the UNDP program (grant for feasibility studies). The "first out" contract model has been used in past projects. (See the discussion of "first out" contracts in Chapter 7, Canada; where Econoler pioneered the concept.)

Since the ESCO that is working in Croatia is a state-owned company, primary attention is not on large profits, but on supporting national interests, mainly energy efficiency and environmental protection. Therefore, the objectives of the company when it was set up were to develop capacity and know-how, find sustainable project financing mechanisms, and develop consumer demand.

Barriers

Consumers show a lack of interest in energy efficiency, probably due to little knowledge in the country about ESCO benefits, and the lack of understanding of the concept in principle. The legislative framework is not particularly supportive of the ESCO concept. Secondary legislation on energy efficiency has not been developed, and the ESCO model is not recognized by the authorities as an individual business model. The result of this situation is that ESCOs cannot invoice their services as a package, and VAT must be paid for the equipment installed for the client, which may jeopardize the profits. Connecting CHP plants to the grid is also difficult. Similar to many other countries, public procurement is complicated.

Education and awareness is increasing about the opportunities offered by ESCOs. More dedicated experts, would be most helpful for energy efficiency project development and implementation in Croatia.

RUSSIA

The following review of ESCO industry development in Russia is provided by Mr. Igor Rokhlikov. His clear understanding of the sequential steps taken to achieve the beginnings of this industry in Russia is very valuable.

The starting point for development of energy saving in Russia was the World Energy Crisis of the 70s and its consequences. Western countries that suffered from the crisis most of all concentrated attention on energy saving problems and this activity was partly moved to Russia. First international programs and pilot projects appeared with the strong support of World Bank (WB), European Bank of Reconstruction and Development (EBRD), and the European Union (EU).

Elaboration of the principles and mechanisms of government policy in energy saving in Russia began in 1992 when the government resolution "On high priority energy saving measures in extraction, production, transportation and use of oil, gas and petrochemicals" was issued and the concept was approved.

In 1995, a decree from the President of the Russian Federation (RF) was issued, "About main directions in energy policy and structure reformation of fuel and energy complex in RF till 2010" and adopted. It highlighted the problem of raising the efficiency of fuel and energy resource uses and "The key issues of Energy Strategy in Russia for the period till 2010" was signed in the same year. These were the documents that formed the basis of legal platform in energy saving.

The federal law on energy saving was issued in 1996 and the first ESCO company was founded later that year. The company founded in Nizhniy Novgorod was a cooperation of three parties: the Regional Administration, "Gasprom" and EBRD. But an effective working structure was not formed because of unclear obligations of participants and the huge scale of the project financed mainly by EBRD.

In 1997, the responsibility for effective use of fuel and energy resources was assigned to the Ministry of Fuel and Energy. The Ministry delegated these functions to regional branches of Gosenegonadzor (the State Energy Monitoring Organization). Their main function was to assign official requirements for machinery and technologies and also to control and monitor implementation. At the same time there was no (and still there is no) state organization in charge of developing the ideas and methodology of energy saving.

The administrative reforms in the Russian government, functions of effective energy use monitoring, were considered in 2004 as surplus for the freedom of entrepreneurship and were closed down. Currently state supervision organizations have no rights to control and supervise energy efficiency. In the end of 90s, the first five national standards in energy saving were issued. They should have become the basis for other standards in energy saving. In 1998 and 2001 two federal purpose-oriented programs were accepted but they were not completely implemented.

In 2003 "Russian Energy Strategy for the period till 2020" was ratified. One of the priority objectives is to raise the effective use of fuel and energy resources and create the essential conditions to move the Russian economy toward energy efficiency. Russian Energy Strategy on the period till 2030 is upon the anvil. Complex government policy in energy saving is only being formed in Russia, but some of its elements are developing quite successfully.

History of Energy Saving in Russia

The first attempt at organizing energy saving in the country was undertaken in a federal purpose-oriented program "Energy Saving in Russia," established by a government resolution in 1998.

The program was meant to gradually lower energy-output ratios of the GDP of the Russian economy in the period 1998-2005. The program focused on energy saving in the fuel and energy complex, housing and public utilities and in power-intensive branches of industry. According to official data the program has accomplished less than 50 percent of its intended results. One of the main reasons for such under-fulfillment was the failure to establish financial mechanism for implementation of energy saving projects. Direct support from the federal budget for three years was around €588,235.3 while, according to the plan, this sum should have been about €75,000,000 (in the prices of 2000).

The second attempt to develop a complex approach to the development of energy saving in Russia was the federal program "Energy Efficiency Economy in 2002-2005 Period and for the Perspective till 2010." The program was to become one of the main mechanisms in the implementation of energy strategy in Russia for the period till 2020. The aim of the program was to build a social-oriented energy economy, that would provide effective energy savings in the country and a sustainable supply. These results were to be achieved by structural reorganization of energy producing and energy consuming industry sectors. The financing of "energy efficiency in consumption" module of the program for the period of its implementation was about 1/5 of the planned amounts. Nevertheless the program helped to raise regional activities in realization of energy saving and energy efficiency policy.

Despite objective preconditions for successful implementation of a comprehensive approach, meaningful results were not achieved because of the following reasons:

- It was impossible to solve energy saving and energy efficiency problems only on federal level. Energy saving consists of dozens of technical, economic and organization actions, that are tied to certain conditions and are interdependent. So their practical implementation can be only on municipal level. The lack of success of the federal program is closely connected with the absence of qualitative municipal programs;
- There is no correlated multilevel system of programs: municipality-region-federation. As the result there are no common policy, goals and tasks;
- There is no effective management system. For some time Energonadzor was in charge of energy saving monitoring, but to externalize management functions is impossible using only state measures;
- Motivation and encouragement of energy saving and energy efficiency were not taken into consideration. Liberal logic's "market will solve all the problems" did not work. As the result energy saving measures were treated as expenditures, but not as measures that allow receiving fringe benefits, economy and profit; and
- Infrastructure of an energy saving market was not built up, starting from the rules regulating the business of the energy service companies (nowadays this term is not officially fixed in any documents), and rules defining the results of energy saving actions. Calculating

the economic effects are not clear. So there is no conductive environment for such a business and there is a lack of the private investments that would be the main force in improving energy efficiency.

ESCO Market Today

The ESCO market in Russia is just forming. The ESCO concept is new and not widely recognized.

There are not many ESCOs and they are not yet ready to provide comprehensive energy saving solutions. ESCO companies, and those who set themselves up as ESCO companies, usually do their business in the region of their location. ESCO activities usually stop after implementation of pilot projects that are partly financed or initiated by international programs. The majority of the ESCO companies are organized on the basis of energy saving equipment production and are working on non-recurring contracts for implementation and service. These contracts do not include long-term programs for providing services to customers. Energy service companies often provide only consulting services and they are not ready to take investment risks. [*Author's note: the use of the term, ESCO, in this section does not necessarily comply with the definition of ESCOs presented in Chapter 1.*]

In Russia there are no statistics on the ESCO business and information on the issue is not sufficient to provide an accurate estimation of the market.

Table 3-1. Statistics on ESCO Companies in Russia

Number of ESCO Companies	Not defined, 4-15 ESCO types and about 100 engineering and consultancy companies
Type of ESCO property	private
ESCO cooperation, alliances	none
Market size	Not defined, potential of energy saving 40-45%

The main ESCO markets in Russia can be divided into public, private and housing sectors.

ESCO—public sector (hospitals, municipal buildings, offices, etc.) characteristics:
1. Constant and measurable energy consumption;

2. Standard technical solutions;
3. Government contracting;
4. Huge potential market;
5. Ability of using performance contracting; and
6. Municipalities are conservative and have lack of experience causing difficulties in marketing, business development and project implementation.

ESCO—private sector (industrial plants) characteristics:
1. Energy consumption is closely connected with technological process and output. This creates difficulties in defining "basic level";
2. Variety and specific character of technological needs;
3. Main approach—to introduce energy management systems; and
4. Often plants do not want to use consultants.

ESCO—housing sector (municipalities, housing cooperatives and apartment houses needing upgrades) characteristics:
1. Improving of heating: insulation, glass units, temperature control systems for the whole building;
2. Projects are often connected with central heating modernization (especially booster stations);
3. Potential ESCO clients cannot pay the whole sum for the services at a time;
4. Hard to work out payment schemes;
5. Technologies are comprehensive, but projects at the household level are inefficient and the norms of dealing with housing cooperatives have not been worked out; and
6. The basis of the work is selling communal services, not performance contracting.

Financing

Sources for financing of energy saving work are:
— Foreign investments, international grants, loans;
— State financing (can provide approximately 10 percent of the sum needed);
— Capital of ESCO companies;
— Capital of customers; and
— Lease financing (a new variant for financing energy saving projects in Russia, used rarely).

In Russia, the use of the traditional and well developed financing mechanisms familiar in other parts of the world is not well developed.

Project Types

The majority of projects are in the public utilities sphere. Their aim is to modernize heating systems and to raise the efficiency of energy consuming equipment. Heating projects are key because Russia annually spends about 30 percent of total energy consumption on heating.

Barriers

Industrial enterprises rarely take energy saving measures in view of low collectability of payments for the consumption of energy resources Financial benefit from the introduction of energy saving technologies is low or is absent. For the industrial enterprises the pressing problems are the development of energy policy and the introduction of effective technologies.

[When Hansen Associates was tasked with assessing the feasibility of performance contracting in Russia by the US government in 1997, it was revealed that a majority of Russian enterprises were not paying their utility bills. The impact on the motivation to use energy more efficiently was effectively summed up by Dr. Yury Kogan of the Krzhizhanovsky Power Engineering Institute, when he commented, "When you are not paying, it makes no difference how much you are not paying."

In addition, the following barriers to establishing an ESCO industry in RF have been noted:
1. Absence of established energy-service companies, capable of ensuring reliable and effective design management, ready to guarantee to investors and users the value of the savings in resources and the periods of investment recovery;
2. Insignificance of the practice of the demonstrative monitoring of the savings of the financial means after the introduction of energy-effective equipment;
3. Absence of rules for the calculation of the obtained savings on a systematic basis, and the accumulation and distribution of data among participants;
4. Absence of effective stimulation: no encouragement for the saving of energy resources, no punishment for irrational energy consumption;

5. Thus far low tariffs on energy. The situation may rapidly change and tariffs will increase;
6. Need for the guarantees of the recovery of investments; and
7. There is no legislation pertaining to ESCOs.

Forecast (Perspectives)

The Ministry of Economic Development is planning to complete the development of the laws, which stimulate shift of the economy to energy-saving and ecologically clean technologies.

One of the versions of this stimulation is recovering of the part of the costs, which enterprises spent on the energy-saving equipment. The requirement of the energy-efficiency of equipment will concern not only industrial enterprises, but also housing and the utilities complex, as well as energy companies, which consume energy for their own needs.

In the next few years, the Russian union of industrialists and entrepreneurs is expected to promote the creation of a national energy service company, which would invest in the introduction of energy-saving technologies. The national energy service company would be occupied not only by consulting enterprises and by development of programs for the implementation of energy-efficient technologies, but would also invest its own capital for implementing these technologies in enterprises, including the delivery of equipment. The energy service company can be established in the form of a joint stock company with the application of a private-state partnership.

In Russia it is possible to form a sustainable ESCO business. There are objective prerequisites for this. The success of an ESCO business will be determined by the possibilities of using the schemes worked out by western companies for financing energy saving projects. Extensive work on the forming of the normative legal basis of this business, the instruction of leaders and specialists should happen soon.

BULGARIA

ESCO History in Bulgaria

Reflecting his expertise in energy efficiency work in Bulgaria, Pavel Manchev has offered the following observations of the ESCO industry and its development.

The ESCO business started relatively late in Bulgaria for several

reasons.
- During the socialists time, energy saving was not a priority. A huge quantity of cheap natural gas and oil were imported from the Soviet Union, so the increased demand was covered extensively.
- For a long period after 1989, energy prices were relatively low. In addition, most of the materials and equipment necessary for EE projects were imported and the financial parameters of potential projects for ESCOs could not support the cost.
- The ESCO model demands a long-term savings stream, and inflation in Bulgaria did not make this possible.

The economic potential for energy savings is believed to be especially high in Bulgaria. Energy intensity is twice that of the EU average, while electricity intensity is outstanding even in the region, four times higher than in Hungary or Turkey. Saving potential is estimated to be up to 50 percent of the energy demand of the building stock, and 30 percent in industry.

There has not been specialized legislation addressing ESCOs. In the current energy efficiency act, there is language dedicated to energy performance contracts, which offers the main rules of the process and the relationships between ESCOs and their partners.

With the support of UNECE and EC, two projects addressing the ESCO methodology were implemented in the years 2003 and 2005.

The first ESCO project was implemented in 1998 by Brunatta in number of municipal building in the city of Rousse. Despite of some problems with retaining the savings in a municipality, the project finished successfully. From 2000, the increasing the prices of energy and equipment financial parameters of projects became better and better. Several municipal projects were implemented fully credit financed (no grants) on relatively taught conditions (interest rate 18 percent and 200 percent collateral). Several projects were implemented mainly in street lighting from producers of SL equipment, which used modified ESCO schemes to attract the municipalities. In the city of Sofia, 350 municipal buildings were included in a project financed by the first big ESCO in Bulgaria, KES JSCo (company for energy services with shareholders RWE Solutions, SWL [Stadtwerke Leipzig] and municipal bank of Sofia). Energy contracts were implemented in several public and residential buildings based on boilers using wood chips.

Another very active and powerful company in recent years is EN-

EMONA JSCo, which even issued two bonds for financing energy efficiency and construction activities.

Current ESCOs Industry and Market

There are about five relatively large and active ESCOs in Bulgaria. There is another 10 companies in the market, practicing forms between EPC and leasing. There is no register supported on national level.

According to Mr. Dichko Prokopiev, chief executive director of ENEMONA JSCo, "The market of ESCO services includes more that one million homes, thousands of public buildings and more than 1000 enterprises. The investment needs of this market are over €6 billion at the moment. This market has to be completely developed in next 12-13 years due the high energy intensity of Bulgarian economy."

Financing and Contracts

Typically, three types of contracts are implemented in Bulgaria

Leasing/investment repayment schemes—most typical examples are ENEMONA JSCo and companies producers/distributors of street lighting equipment. They finance and implement the projects. The client repay on the bases of monthly instalments usually for period of 7 years

Shared savings—implemented by KES JSCo and Brunatta in municipal buildings—the cities of Sofia and Rousse

Energy supply—practiced mainly by producers and distributors of heating equipment. For instance ERATO Holding JSCo installs its own production biomass boilers and respective accessories, delivers chips. The client pays 30 percent lower price per MWh than the price of MWH produced by naphtha or electricity.

In all cases, ESCOs have used their own financial means or credits. There are a lot of financial institutions ready to finance ESCO projects. The first of all specialized credit lines of EBRD, which include and grant component provided by Kozloduy International Ecommissioning Support Fund (KIDSF). Since 2006 the Bulgarian Energy Efficiency Fund (BEEF) is operational, providing financing for ESCOs as well, but in 2006 only one ESCO applied. The BEEF offers three types of help: partial credit guarantee, joint crediting with commercial banks and technical assistance for project development.

The use of partial credit guarantees in Bulgaria is supported by USAID. Working with multi-lateral banks, the guarantees are designed to increase the availability and access to credit for municipalities through a commercial bank. In 2005, the available loan was $15 million, 13 percent of which was used for energy infrastructure improvement, including ESCO investments. The Facility for Municipal Energy Efficiency under the USAID is more than just financial help, as technical support is also provided for clients.

Most commercial banks offer credits for energy efficiency or are involved in servicing of credit lines financing energy efficiency.

The EBRD is supporting the development of the market in Bulgaria to promote energy efficiency, with a loan to the Bulgarian ESCO Fund— Energetics and Energy Savings Fund SPV (EESF), a special purpose company to finance the energy services business of Enemona AD, a construction and engineering group.

Typically the fund has supported energy efficiency projects in kindergartens, schools, hospitals and other public buildings.

The €7 million loan, to be provided in two branches of €3 million and €4 million, will help provide Enemona with the long-term capital it needs to expand its ESCO business in Bulgaria. The proceeds will be used to purchase receivables from energy performance contracts carried out by Enemona, the Fund's majority shareholder.

Since its establishment in 2006, EESF has purchased receivables from more than 20 energy performance contracts (EPC) with a total nominal value exceeding BGN 8 million.

Types of ESCO Projects

Most projects address buildings and street lighting. Only one project is implemented in district heating.

Major Market Segments

In most cases recipients are municipalities. There are two energy supply projects implemented in residential buildings in Sofia.

Barriers

The fact that ESCOs do not sell mere products, but more complex services and promises of future energy and financial savings, makes their

operations that much more complex than regular businesses. Hence, besides the usual challenges of running any business, ESCOs have to face the following barriers:
- The lack of awareness and knowledge in implementation of the scheme on both sides;
- Difficulties coming from baseline and energy savings determination (in most of cases it is necessary to be normalized, no national M&V);
- Problems with retaining the savings by municipalities;
- Lack of trust between partners—municipalities have suspicions regarding the negotiated results and ESCOs are afraid about regularity of payments;
- Clients, especially industrials, are intimidated by the idea of handing their energy management over to an external company;
- Given the thin financing capacity of the ESCOs, third-party financing is often necessary. However, the concept of third party financing is misunderstood and not trusted enough by the potential ESCO clients; and
- The uncertainty of ESCO regulation/legislation presents an additional risk to investors. While the definition of ESCOs as any other commercial company is certain, the peripheral laws may hamper their profitability have historically not been stable—such laws related to ESCOs were first adopted in 1999, abolished in 2001, and now expected to be reinstated in 2003.

Enabling Factors

The very new positive opportunity exists in the language of the energy performance contracting section of the new draft of the Energy Efficiency Act. This article envisages state or municipal property as subject of EPC in the budgets of respective ministries or municipalities and allows planned and ensured financial means for the execution of the contract to correspond to the normalized expenses for energy of these buildings.

The perspectives are optimistic due several reasons:
- All state and municipal buildings with heating area over 1000 m^2 have to be certified till year 2012;
- Financing will be ensured due obligations of Bulgaria as EU member state;
- A big demand for companies to implement the measures is expected;

- A similar situation could be expected in residential buildings sector; and
- Industrial sector has to cover emissions obligations (according to the scheme for quota trading) which is impossible without implementation of energy efficiency measures, and may use ESCO scheme to solve the problem.

ESCOs could use credit money from the EBRD credit line with 15 percent grant on principal. Bulgarian Energy Efficiency Fund (BEEF) offers "soft" loans. The fund has a guaranty tool which could provide guarantee up to 50 percent of credit. Both the EBRD credit line and BEEF provide technical assistance for preparation of EE projects and financial schemes.

Chapter 4
Africa

Africa is a very large continent with varied energy needs. For example, Côte d'Ivoire is a net exporter of energy while South Africa is struggling with a capacity shortage.

This chapter offers a sampling of ESCO development across the continent. One section looks at northern Africa with information about Algeria, Morocco and Tunisia. From western Africa, a look at the ESCO industry in Côte d'Ivoire contrasts with an overview of the status of performance contracting in Kenya in eastern Africa. The chapter concludes with a more in-depth look at South Africa.

The authors are very pleased with the caliber of contributing authors, who have taken the time to share their insights on the ESCO industry in their respective parts of Africa. Mr. Hakim Zahar provides an excellent overview of the energy picture in Algeria, Morocco, and Tunisia and its significance to the developing ESCO industry in northern Africa. Dr. R. M'Gbra N'Guessan of the West African office of Econoler offers a clear picture of the ESCO opportunities and obstacles in that part of the continent through an assessment of the industry in Côte d'Ivore. Traveling east, Mr. Paul Kirai provides special insights into activities in Kenya. Finally, Dr. L.J. Grobler and Mr. A.Z. Dalgleish give us an interesting look at a country troubled by capacity problems and how that affects ESCO growth.

NORTHERN AFRICA

The potential for energy efficiency measures in northern African countries is estimated to be more than USD 1.5 billion annually over the next ten years. This corresponds to three to five percent of the total costs of energy consumption in the three countries considered: Algeria, Morocco and Tunisia. The greenhouse gas emission reduction potential is also tremendous.

The situation varies from country to country. While Morocco and Tunisia have faced large deficits in terms of energy supply vs. demand since the end of the 90s, Algeria is one of the largest natural gas exporters in Africa, and in the world. As a result, the energy efficiency view is structured differently from one country to the others. Tunisia was the first country in northern Africa to develop a policy. It created an energy conservation agency "L'Agence Nationale pour la Matrise de l'Énergie" (ANME) in 1985.

In 2006, Morocco proposed the creation of an energy efficiency policy and organized the administrative structure in the EE field. The renewable sector benefits from the extensive experience of the "Centre de Développement des Energies Renouvelables (CDER)."

Algeria developed its energy efficiency law in 2006, but is struggling to enforce it despite the existence of the "Agence de Promotion et de Rationalisation de L'Utilisation de l'Énergie."

The support provided by the three governments differs, too. While Morocco and Algeria consider that the development of energy efficiency needs to be on a commercial basis only to ensure its sustainability. Tunisia understood that the field has several barriers, and without an adequate incentive it will not be possible to develop the market.

The situation of each country leads to a different perception of the ESCO market, and the trust in its achievements is very different as well. The first tentative creation of an ESCO was realized in Morocco in 1993. Econoler International invested to create ADS-Maroc in 1993 to develop the energy efficiency market and propose, where possible, the ESCO concept. At that time, the awareness of the main stakeholders and the utilities was very low. The resistance of the market required a review of the ESCO concept and it was limited to a specialized energy efficiency consulting group, which could realize more than 40 projects within three years of its start-up. At the same time, several other consulting engineers were also emerging under a USAID program and help from other donors. As of today, Morocco has not yet incubated any real ESCO.

The level of energy pricing and the maturity of Algerian Industries in the 90s is certainly among the reasons why Hydro-Quebec from Canada was interested in entering into a partnership with the "Société Nationale de l'Électricité et du gaz, SONELGAZ" in 1996. However, the project could not be completed despite a promising business plan. Some administrative difficulties could not be overcome and today there is no ESCO operating in Algeria.

Energy efficiency programs and initiatives in Tunisia are being channeled through the ANME under the Tunisian Ministry of Industry and Energy (MOIE). In 1998, the ANME was transferred to the Ministry of

Environment and Territory and ANME's activities, especially in the industrial sector, have subsequently been reduced. With a 2001 reorganization, the responsibility for energy efficiency measures in all sectors, especially the industrial sector, was reassigned to the MOIE as a specific division of the Ministry's Energy Department. This reorganization was completed in September 2002 when the ANME was retransferred to MOIE's authority.

In view of these changes, the "Société Tunisienne de l'Électricité et du Gaz, STEG," the Tunisian utility and Hydro-Québec Canada decided to invest in the first ESCO in Tunisia. Two Tunisian banks were also interested and the 'Société Tunisienne de Gestion de l'Énergie, STGE" was created in 1999. The STGE operated as the unique ESCO until 2005 when the energy efficiency program for the industrial sector, funded by World Bank and GEF decided to take ESCO market development under their wing. Today, the Tunisian market includes at least four ESCOs registered by the ANME.

Barriers to ESCOs in Tunisia

The Tunisian ESCO market is still suffering from several barriers. The main barriers are as follows:
1. Lack of financing for energy efficiency investments.
2. Energy efficiency investments are a new type of activity, as the return on investment is based upon operating cost savings and not on increased revenue.
3. Inadequate information available in the industrial sector.
4. Industrial end-users are more concerned with enhancing operations through improved production and productivity rather than with reducing operational costs, including through energy efficiency measures.
5. Lack of expertise and intermediaries who could develop projects. The existing ESCOs are derived from consulting firms and have limited knowledge of the measurement and verification protocols.

The institutional framework for energy efficiency projects has been improved by the GEF project in the industrial sector. However insufficient technical and financial tools and high administrative barriers have hindered the starting up of an energy efficiency market. Since 1985, the audit program has supported 50 percent of audit costs. A large number of auditors are operating and targeting the market. The quality of the audits was limited and until 2004, only four percent of the audits could be converted into real projects. This historic handicap created a bad press for energy efficiency and the ESCO market.

One of the main barriers to energy efficiency measures in Tunisia is the lack of project financing on reasonable terms resulting from commercial financial institutions' unfamiliarity with assessing energy efficiency investments. Moreover, the production priority bias in the industrial sector, together with a lack of information on the benefits of energy efficiency, have thus far constituted insurmountable barriers to the implementation of energy efficiency measures. This, despite the presence of ESCOs in the market and mandatory energy audits for large industry.

In order to attract the attention of end-users it is, therefore, key to provide a small financial incentive to entice them to tackle the issue of energy efficiency and realise the potential in reduction of energy consumption. The failure of the mandatory energy efficiency audits in bringing about investments is due to the fact that they don't come along with a suggested financing package, and are administered by an organisation—ANME- that is perceived to be too academic.

The GEF pilot phase objective is to help demonstrate that shorter paybacks, financial sustainability and replication, can be achieved through the ESCO model and, in particular, that market aggregation for the bulk-purchase of the EC measures as an integral part of a package of efficiency services will lead to economy and efficiency. The phase also provided an additional 10 percent when the projects were implemented under the ESCO model without limitation as to a shared or guaranteed saving model of contract.

A partial risk guarantee fund was also created and was intended to address the financing barrier through the provision of guarantee-backed financing for energy efficiency projects. In the context of the partial risk guarantee facility, the risk of the loan is shared between the commercial bank and the ESCO. The latter is in turn backed, in part or totally, by the guarantee facility at 50 percent of the total loan. This corresponds to other guarantee funds, such as the Hungarian Energy Efficiency Co-Financing Programme (HEECP), which also provides a guarantee of 50 percent of total bank exposure.

The guarantee facility will also assist in the creation of ESCOs, thus addressing the lack of intermediaries who could develop projects. The ESCOs will be key actors to guarantee a sustainable and sustained market. The fund is expected to have a sustained market transformation effect by lowering the perceptions of risk on the part of commercial banks and end-users regarding energy performance contracting and end-user financing models.

Major Market Segments and Types of ESCO Projects

The energy efficiency market in Tunisia is characterized by a large potential in the industrial and commercial sectors. The Tunisian industrial sector is composed of SMEs with few large industries. The ESCO projects for the SMEs have rather low investment size (less than USD 300,000). The industries are mostly private and decisions are generally centralized at the managerial level.

The Tunisian commercial sector is characterized by private entities in tourism, the banks and other administrative headquarters, and some few private clinics and commerce. The size of the EE project is also small and the investment is usually less than USD 250,000.

Tunisia has been trying to develop the use of cogeneration since 1996. This technology has suffered from a weak institutional framework. In 2006, the energy efficiency law was developed for cogeneration. The STEG has the obligation to buy the excess electricity generated. The tariff, however, is still low (around USD 0.05) and the spread between the electricity cost and natural gas does not yet allow for easy development of these projects.

Standard new efficient equipment (like variable speed drives, efficient lighting, controls, etc.) combined with thermal efficient measures, constitute most of the technologies offered by the Tunisian ESCOs. The main energy efficiency technologies that are likely to be utilized are energy-efficient burners and boilers, variable speed drives (VSDs), high efficiency motors, power factor improvements, compressors, controls, steam traps, and heat recovery devices.

Positive Enabling Factors

In 2007, Morocco developed large DSM programs, including auditing and other incentives for the development of the ESCO market. However, this development is relatively recent, so it is difficult to predict if the ESCO market will react positively to this new strategy. Electricity tariffs have not increased over the past five years and this has sent bad signals to the market. The Moroccan Government will have to consider increases in fuel and the natural gas costs. It is expected that the tariffs will most likely increase in the coming years. This factor is positive for the ESCO market development.

Tunisia is heavily supporting ESCO market development. More than 23 percent of the investments are provided as a grant on ESCO projects. In addition, the guarantee fund covers more than 75 percent of the funds provided by the ESCO. With the recent creation of an Energy Efficiency Fund, the signal provided to the market will certainly attract the large ESCOs to build strategies for Tunisia.

Outlook

Northern Africa is characterized by a historic capacity for negotiation that may kill ESCO market development. Several end users, even with high educational levels, become "carpet dealers" when treating with ESCOs. The maturity of the EE market is still to be built. The market will remain thin for the coming three to five years. This fact may destroy even the efforts done by the existing ESCOs to build a reliable and sustainable market. The governmental sector may be the only solution for the development of the ESCO market based on guaranteed savings. When this step will be crossed, the private sector will copy its big brother.

CÔTE D'IVOIRE

Energy efficiency project development was introduced in West Africa during the period 1980-1990, particularly in Côte d'Ivoire and Senegal through a French cooperation initiative involving the national electricity companies. In 1993, the Global Environment Facility (GEF) approved a regional energy efficiency pilot project implemented by UNDP, enabling the two above-mentioned countries to determine the energy efficiency potential, both in public and private buildings. Despite the great potential for energy savings in these two countries and the positive results of pilot projects, no important investment was made during the project lifetime, or thereafter, to support the development of the EE market in the region.

Côte d'Ivoire is an emerging economy in Sub-Saharan Africa. Since energy needs are growing at a fast pace, hydropower can no longer provide the country with energy resources needed to meet the demand. The government in Côte d'Ivoire has adopted a new energy policy plan based on fossil fuels reduction, since they generate CO_2 emissions. Energy efficiency initiatives were not previously emphasized in most industrial and commercial sectors of the country. Lessons learned from previous energy efficiency projects showed that new approaches, such as ESCOs, are needed to extend energy efficiency services into the industrial and commercial sectors.

Energy efficiency consciousness was introduced in the country twenty years ago through the action of the Energy Efficiency Office. A few consumers also realized successful advances in implementing energy efficiency measures. A hotel in Abidjan used heat generated by its air conditioning systems to produce hot water. Capacitors have been installed at many factories. UNDP has completed energy audits of the major buildings in the country. Technical knowledge has been developed, paving the way for the introduction of the ESCO concept.

Barriers

The two main obstacles have been the lack of financing and the absence of qualified entrepreneurs in the energy efficiency field, which could support the implementation of the performance contracting approach, or third-party investment. Taking into account this context, the Institut de l'energie et de l'environnement de la Francophonie (IEPF) and the World Bank agreed in 2000, to launch a pilot programme to develop the ESCO market in Côte d'Ivoire in order to overcome the obstacles to energy efficiency investment in the region. The objective of the Côte d'Ivoire ESCO Project was to establish a sustainable commercial market for energy efficiency services between suppliers and end-users, obtain energy savings, and the associated reduction in greenhouse gas emissions. The project implementation started in June 2000 with capacity building activities. It has successfully engaged four companies in developing the ESCO business in Côte d'Ivoire: two large companies with an existing client portfolio to which they provide maintenance support, and two individual companies dedicated to offering energy efficiency services.

Current ESCO Industry and Market

To achieve its objectives, the ESCO project in Côte d'Ivoire provided various results that contributed to energy efficiency market development at the national level:
  (i) A reliable service supply was developed;
  (ii) Energy efficiency service demand was developed; and
  (iii) Financial resources were made available at the National Investment Bank as a guarantee fund to support the ESCO investment.

The current ESCO industry in Côte d'Ivoire is dominated by three of the four enterprises initially supported by the IEPF Project. All of the

ESCOs are engineering companies, but one of them is also an equipment supplier. They are now offering their services both to the commercial and industrial sectors as well as to institutions. All of them are profitable even though the political and social crisis has affected the whole economy. More than 20 new performance contracts with ESCO clients have been signed during the last three years. These projects have resulted in positive impacts on client energy bills.

The ESCO industry in this country is now expanding after five years of socio-political crisis. The ESCOs are active on the regional energy services market, mainly in Benin, Burkina Faso, Mali, Senegal and Togo. New engineering companies are aware of energy efficiency advantages and they have included such services in their core businesses. With increasing fuel prices, demand for energy efficiency services will rise even further in the future.

High-level government support has been granted for the ESCOs. By focusing on no cost/low cost measures with short payback periods, the strategy adopted by EE services providers seems appropriate for the emerging energy efficiency market in Côte d'Ivoire. The market actors have demonstrated their ability to respond to difficulties that arise, for example, the leveraging of financial resources to ensure adequate investments in a risky environment.

Financing

The market appears to have addressed all the relevant issues in a holistic approach by developing demand and supply as well as financing aspects. However, the market needs to have more involvement with local financing institutions to enlarge the opportunity to exploit the potential for energy efficiency in industry.

The ESCO market development in Côte d'Ivoire has called for seed capital to demonstrate the feasibility of the concept in a new environment. Resources have been provided for ESCOs' initial needs during the time when they do not collect revenue because of their training and commercial campaign periods. Each of them received USD 10,000 in seed capital for equipment, living allowances and initial overheads.

The provided resources have been allocated as a no-interest loan and reimbursed as a fixed portion of their future revenues. The allocation is specific to each ESCO and it was submitted for no objection by the IEPF to the steering committee.

IEPF which acts as the ESCO project executing agency set up a revolving fund (RF) that amounts to USD 200,000. The RF has been administered by a local microfinance institution based in Abidjan. This fund has been used: (i) to finance, partially or totally, the end-user's investment proposed by an ESCO under a performance contract, and (ii) to guarantee, partially or totally, a loan by a private commercial institution to an energy end-user under a contract with an ESCO.

The Credit Mutuel de Côte d'Ivoire (CMCI) is the RF administrator. It has operated the RF with practices and rules similar to commercial institution loans or credit lines. An 8 percent interest rate was applied to the first year and it reached 16 percent in the third year. These conditions have progressively made it tougher to reach the usual commercial standards.

During the ESCO market development project, emphasis was put on mobilizing local financing institutions that could provide additional resources to promote energy efficiency businesses in Côte d'Ivoire. The project partners provided the commercial banking system and other financial institutions with information on an ESCO's approach and on the energy efficiency market. Special attention was paid to supporting ESCOs in their search for private funding and to guide their innovative methods to mitigate the financial risks.

After completion of the ESCO market development project, the remaining resources in the revolving fund have been used by the Ivorian National Investment Bank (BNI) to set-up a guarantee fund, which aims to develop co-operation with local commercial banks to find sustainable financial mechanisms that will carry on beyond the project.

Contracts

Various types of contracts are offered by the ESCO industry. An adequate model should be selected for successful implementation of energy efficiency projects, taking into account the local legal context. The most common ones used in Côte d'Ivoire are the "Guaranteed Savings" and the "Shared Savings" contracts. The quotation marks and capitals are used here to highlight the fact that the models used in Côte d'Ivoire are not the same as those defined in Chapter 1. The Côte d'Ivoire models are defined below.

Guaranteed Savings. In a Ivorian guaranteed savings contract, the ESCO is remunerated for the whole of its services and it guarantees annual savings to the client. The ESCO contract in this context guarantees to the client to save a certain amount of energy. Under this contract, the ESCO usually includes the design, installation and savings performance risks, but does not assume credit risk of repayment by the customer.

Shared Savings. In a shared savings contract, the ESCO is remunerated according to a percentage of energy cost savings, varying from 10 to 90%, according to the nature of the investments and the terms of contract. The investments are generally carried out by the ESCO. The cost savings are shared between the ESCO and the client for a certain contract period in accordance with a pre-arranged share. In this contract model the costs, contract period and other risks are shared between ESCO and the client. The shared savings contract includes performance and credit risks.

Types of Projects

ESCOs in Côte d'Ivoire are offering a complete turnkey package of energy efficiency services including energy auditing, feasibility studies, brokering of financing, specialized documentation for performance contracting, project engineering and implementation along with monitoring and verification services. The Table 4-1 illustrates the type of projects undertaken by ESCOs after an energy audit had been completed in existing facilities. The source of financing is shown in the last column.

Obstacles to ESCOs

Previous activities undertaken by the government for the promotion of energy efficiency in Côte d'Ivoire have faced various obstacles. The IEPF project was designed to overcome market obstacles to commercialization of energy efficiency services in the private sector.

End-users are becoming increasingly aware of the potential profits that could arise from a more efficient use of energy. However, even in relatively big companies, the internal skills have not been available to promote energy efficiency. Consequently, energy efficiency projects have not been considered high priority in the management agenda. Investment resources have been scarce, which fosters competition among various types of internal projects. Traditionally, conservative commercial banks do not show much interest in energy efficiency projects as they do not understand them.

External proposals have been perceived either as not reliable or too risky. The end-users tend to believe that they are bearing the whole risk of the "innovative" approach while the proposers are paid for their work

Table 4-1. Recent Projects Portfolio

ESCO	Clients	Type of projects	Estimated Energy Saving KWh/yr	FCFA/yr	Impact CO2 T/yr	Source of financing
ECA	2. Universite Abobo	Education Sector	170 000	14 060 000	47	Revolving Fund
	3. Postel 2001	Office Sector	2 200 000	98 000 000	604	Client financing
	4. CHU Cocody	Health Sector	0	0		Revolving Fund
	6. BICICI Tour	Bank Sector	0	16 161 039	0	Client financing
	7. BAD	Bank Sector	4 322 000	215 600 000	1 186	Client financing
	8. Hotel President	Hotel Sector				Bank
ISE	1. SIM Ivoiris	Commercial	0	4 882 000	0	Client financing
	2. SIM Ivoiris (headquarter)	Commercial	656 320	28 405 492	180	Revolving Fund
	3. CI Telecom	Commercial	0	60 000 000	0	Client financing
	4. CI Telecom République	Commercial	164 078	10 525 077	45	Revolving Fund
	5. Direction de l'Energie	Commercial	0	11 400 000	0	Revolving Fund
	6. CI Telecom KM4	Commercial	751 140	30 763 396	206	Revolving Fund
	7. CI Telecom Banco	Commercial	269 000	16 169 483	74	Revolving Fund
	8. CI Telecom Adjamé	Commercial	356 808	20 329 950	98	Revolving Fund
	9. Orange (Imm. Lumière)	Commercial	323 541	21 634 238	89	Revolving Fund
	10. Orange (Imm. Saha)	Commercial	143 740	7 200 000	39	Revolving Fund
	11. Orange	Commercial	0	19 256 243	0	Client financing
	12. SODECI	Commercial	0	240 000 000	0	Client financing
COGIM	1. Blohorn (1)	Industry	36 200	2 025 000	10	Client financing
	2. Blohorn (2)	Industry	45 000	667 000	12	Client financing
	3. Shell	Industry	905 000	16 420 000	248	Client financing
	4. Premoto	Commercial	52 000	5 455 000	14	Revolving Fund
	5. Les Hévéas (1)	Commercial	269 000	12 720 000	74	Revolving Fund
	6. Les Hévéas (2)	Commercial	269 000	12 720 000	74	Revolving Fund
	9. Palm-CI Ehania	Industry	31 300	89 690 000	9	Revolving Fund
SEEE	1. BAD	Bank	1 358 000	74 800 000	373	Client financing
	2. AERIA	Airport	3 182 000	130 428 000	873	Client financing
	3. SGBCI Siège	Bank	305 000	26 800 000	84	Revolving Fund
	4. SGBCI Vridi	Bank	138 700	7 540 000	38	Client financing
	5. Club Med Assinie	Hotel	10 000	13 867 600	3	Client financing

whatever the performance result.

The electricity tariffs, which amount to eight to ten cents per kWh for medium voltage users, do not constitute a drawback as they are realistic enough to ensure the power sector development and energy efficiency project profitability.

KENYA

The following review of the conditions in Kenya, which have historically hindered/helped ESCO development, has been provided by Paul Kirai. Drawing upon his experience as an energy consultant in Kenya, He provides an excellent overview of the problems faced by ESCOs trying to establish a new industry.

Short History of the ESCO Industry

During the 80s, the government, through the Kenyan Ministry of Energy, made an attempt to introduce energy efficiency services. The government officers gave free energy audit services to industry, but there was little implementation. Neither was there any private sector involvement in the services. The government-owned utility also started a DSM programme, but there was no incentive to see it through since the utility was not profit motivated. As a result, these programmes did not have much impact.

In 2001, the GEF/UNDP supported energy efficiency project revived energy efficiency services in Kenya in a more robust manner. This project increased awareness, developed capacity, engaged financial institutions, initiated policy reviews to include energy efficiency and generally developed the energy efficiency market. Even with these developments, no viable ESCO emerged, as was hoped, at the end of the project.

Previous attempts to introduce ESCOs were led by "marketers," who offered quick fix solutions to unsuspecting clients. These solutions were mainly in the form of "black boxes" of ambiguous nature and exaggerated performance, which were supposed to save up to 30 percent of electricity consumed. Most of the clients, who bought in, lost their money and were dissatisfied by the services and products offered by these companies. This may have resulted in a "negative demand" for energy services.

The Kenyan ESCO Development Process

In 2005, the GEF-KAM Energy Efficiency Project resolved to establish an ESCO in Kenya by the end of the project in 2006. This was to be done using the trained manpower within the project and by providing training and institutional support through the Kenya Association of Manufacturers.

In 2005 Econoler International was engaged to help develop a business plan for the ESCO. The plan provided an evaluation of the market size, the market approach, the business process to be adopted and the financial results to be expected from this new operation. A training programme was conducted for the new team by Econoler, which also included a study tour in Canada. A set of basic ESCO tools was developed.

In early 2006, a Team consisting of former GEF-KAM Project engineers established "Integrated Energy Solutions," an energy service company, as the first ESCO in Kenya. The new company was dedicated to the realization of energy efficiency projects for the benefit of its clients on a performance-based contract approach.

The ESCO had the following objectives:

- Develop and implement energy efficiency projects in Kenya, by addressing the different barriers related to the implementation of such projects—particularly financial and confidence barriers;

- Apply new innovative financing mechanisms for energy efficiency;

- Introduce professionalism in the provision of energy efficiency services in Kenya; and

- Be a sustainable, profitable operation, and meet the objectives of its different shareholders.

At the same time, a small part of the GEF-KAM Project budget meant for demonstration projects had been channelled to KAM through a tripartite agreement involving KAM-UNDP and UNOPS to catalyze ESCO activities. Through a competitive bidding process, KAM contracted a young ESCO to test and prove the ESCO approach by developing and implementing "ESCO type" projects over a two year period. The ESCO would be paid for the development services, and it would also keep any revenues generated from the actual projects developed and implemented. The contract fees would only cover minimum operating costs but would not be sufficient for purchase of equipment or even investments in project implementation. It was expected that the income generated from the implementation of projects would be sufficient to capitalize the ESCO after the contract with KAM came to an end.

Current ESCO Industry and Market

So far there are no other known ESCOs in Kenya. A number of engineering firms have started offering energy efficiency services since the GEF-KAM project. Many of these firms benefited from training under GEF-KAM project. However, none of these are offering performance-based contracting. Also, the last two years have seen a number of new players enter the market in energy equipment supply – particularly lighting solutions and power regulators.

Financing and Contracts

The strategy for the new company was to make the offer attractive to clients. The ESCO would initially offer direct project financing to its clients with only a small portion of co-financing. The ESCO aggressively marketed the shared savings approach and guaranteed savings. Shared savings was preferred as it would address the financial barrier identified by industry. For this to succeed, it was important that the ESCO identify a source of concessionary finance in the first round of projects. However, this did not happen and even though the ESCO identified suitable projects, it did not have the capital to invest.

The ESCO has been in discussion with the clients to finance those projects, but it has not been easy since the ESCO cannot offer any fiscal guarantee. This has prompted a modification of the approach – where the ESCO is charging near market fees for development of the project as well as fees for "hand-holding" the client through the implementation process, in exchange for a "bonus" payment at agreed intervals based on verified energy savings realized. This approach is receiving more favourable attention as most clients seem to understand the concept of paying for services and a bonus if "expectations are met."

Major Market Segments

At the beginning, the ESCO adopted the following marketing strategy:
- Start with clients who have benefited from the UNDP/GEF KAM energy project, who are already aware of the benefits of energy efficiency projects;

- Focus on clients with projects that have short payback periods, of no more than 1.5 years; and

- Focus on clients that could have an influence on the market because of their "leader image," financial stability, etc., to develop some highly credible references in the market.

As a result, the ESCO has largely concentrated on the private sector. The greatest focus has been on the manufacturing industry, followed by hotels and hospitals. There has been interest in commercial buildings, especially for lighting solutions, but this has not translated into any real projects.

Barriers to ESCOs and EPC

A combination of poor understanding of the energy efficiency market and the lack of confidence among the energy experts and the clients continues to be a major barrier. Many clients are unwilling to accept energy efficiency projects with payback periods of more than two to three years even though they can accept longer periods for other types of projects.

ESCOs are not capitalized to finance projects. Kenyan financial institutions are unfamiliar with the concept of energy efficiency financing; hence, no funds are available to support ESCO-type projects. A survey of Kenyan financial institutions established that, although there was willingness to explore new business opportunities, very few of them were willing to change traditional lending practices based on the balance sheet of the client. Few would consider lending to the ESCO on presentation of a bankable energy project.

Any commercial loan is "on balance sheet."

Many experts have recommended the establishment of an EE fund (including third-party financing) to finance ESCOs or their clients, and stimulate the market for EE services.

Another potential barrier has to do with clients paying for verified savings. The nature of the ESCO contract requires an open accounting system so that the savings can be tracked in the financial system of the client. This is not always the case. Therefore contracts need to be less complicated and easily verifiable. Disputes can lead to protracted legal processes with uncertain outcome.

Other barriers include:
- Lack of previous local case studies and absence of successful ESCOs in Africa; and
- Little interest by the utilities in improving energy efficiency and therefore providing backing to the ESCO.

Positive Enabling Factors

The Kenyan economy is showing signs of recovery. In 2006 it posted a real growth of 7 percent compared to 4 percent in 2005 and 1.2 percent in

2002. Other favourable factors include:
- Publishing of the energy act in 2006 that expressly recognizes the role of ESCOs in implementing energy efficiency measures. The act also requires certain facilities within a certain threshold energy consumption to carry out annual energy audits and to submit reports to the Energy Regulatory Commission;

- Bank interest rates reduced as the government domestic borrowing dropped; and

- Growth in electricity demand surpassed the projections of 6 percent by the year 2014. By 2007, it had already risen 7 percent.

Due to the nature of the market and the many barriers identified, ESCO associations have played a key role in promoting ESCO services in many countries. Given the relatively small size of the Kenyan market, it may be too early to form an ESCO association. However, support for the establishment of a regional ESCO association would be useful. A regional ESCO association would support peer exchanges among stakeholders in the region to share experiences with EE project financing and implementation. It would also increase exchanges between ESCOs of the region and with other countries where ESCOs are well established, helping to accelerate the maturing of the ESCO industry over the entire region.

Outlook for the ESCO Market

A number of factors seem to converge towards a greater need for ESCO services, including:
- High energy prices, recent upward tariff review of electricity together with fuel cost adjustments have resulted in a 50-80 percent increase in power costs;
- The government energy policy, the Energy Act 2006, and rising energy prices, are expected to prompt growth in energy demand and energy saving potential; and
- Growing awareness of the causes and impact of climate change.

Brief Case Studies of Potential ESCO Projects in Kenya

The single Kenyan ESCO has carried out numerous energy audits (2007-2008) from which the following potential projects have emerged. They are currently under development or negotiation.

Table 4-2. Potential Projects in Kenya

Company	Energy saving measures	Quantity of saved energy	Cost of saved energy KSh	Investment required KSh	Comments
Battery Manufacturer	a) Combustion the air/fuel intake efficiency improvement by introducing a semi-automated burner to regulate the air/fuel intake	20,000 litres of fuel oil per month	Savings on FO per month is kshs 939,227 and annual savings equals kshs11,270,725.1	Introduction of burner services, locked air to fuel ratio at cost of kshs 4,209,000	Simple payback period 4.48 months
	b) Tariff change by change of power supply from medium to bulk supply		Savings per month is kshs 300,000 kshs 3,600,000 per annum	4,000,000	Simple payback is 13 months
Cement Company	Recovery of the waste heat at preheater exhaust to be used to generate power	Heat convertible into power is 2032 kW which is 15% of the plant requirement	Annual cost of savings kshs 85,568,911	Investment for equipment only is kshs 450,000,000	Simple payback is 63 months
Private Hospital	Tariff switch from B1 to C2 at KPLC supply	11% of total power consumed	Kshs 90,000 per month KSh 1,080,000 per annum	Kshs 400,000 for transformer installation	Simple payback is 4.42 months
Milk Plant	Switching from industrial diesel oil (IDO) to fuel oil in one of the plant boilers	Switching from 12,000 litres of IDO to 12,000 litres of FO per month	Savings per month is kshs 300,000 or 3,600,000 per annum	Installation of a new burner that can combust FO in the boiler costing around kshs 1,500,000	Projected payback period is 15 months
University	a) Use of efficient technologies in lighting	10% of electricity consumption will be saved	Potential for reduction in annual electricity bill from kshs 120M to kshs 108M hence savings of kshs 12M per annum	Investment in various efficient technologies will cost about kshs15M	Simple payback period is 5 months
	b) Fuel substitution for cooking-electricity to gas or gas to fuel oil for water heating				

USD = 63KSh

SOUTH AFRICA

The history of the ESCO industry development in South Africa, and a review of the current status of the industry is very ably presented by Dr. L.J. Grobler and Mr. Braam Dagleish.

Eskom, South Africa's national electric utility, formally recognised demand side management (DSM) in 1992 when integrated electricity planning (IEP) was first introduced. Recognizing that South Africa might run out of capacity by 2006, the first DSM plan was produced in 1994. In this plan, the role of DSM was established and a wide range of DSM opportunities and alternatives available to Eskom were identified.

From small beginnings in the early 90s, starting with research, pilot studies and time of use tariffs, Eskom's DSM programme has grown into a concerted national electricity-saving effort officially initiated in the last quarter of 2002.

The 1999 launch of the local efficient lighting initiative called Bonesa was among the major milestones in the early phase of DSM in South Africa. This was jointly funded by the Global Environment Facility and Eskom over a period of three years. This promoted the use of compact fluorescent lamps (CFLs) through customer education, advertising, handout campaigns, and marketing. The focus was on lowering the price of energy-efficient lamps. The CFLs were originally priced between R60 and R80 per lamp. In 2004, the price of CFLs dropped to between R13 and R20, due to joint sales promotions with local suppliers and increased sales volume of CFLs. The annual review of residential lamp sales shows that CFL sales contribute 15 percent towards the total South African market with a current retail price of R10 per unit.

Early on, Eskom viewed the use of the energy service company model as an effective DSM tool. In the mid-90s, Hansen Associates conducted two weeks of intensive ESCO training for the utility. In September 2002, the DSM Fund was approved, and 2003 was spent mainly on setting up the DSM business model and operations, customer awareness and education campaigns, as well as the establishment of an energy service company (ESCO) industry.

During 2004 and 2005, demand side management started to pick up momentum, with clients and projects being signed and implemented in all its market sectors.

Following the major electricity blackouts experienced during the first quarter of 2006 in the Western Cape, an Eskom Energy Crisis Committee (regional) was established. The committee was responsible for the complete strategy and activities to alleviate the energy constraints experienced in the Western Cape.

Eskom and the Department of Minerals and Energy (DME) recognized that success in the Western Cape could be replicated on a national scale, which would help South Africa confront an expected power crunch for the next three to four years or more.

The National Energy Efficiency Agency (NEEA) was established in 2006 with a broad mandate to promote energy efficiency throughout the South African economy. This agency is charged with supporting energy efficiency projects for the public sector and targeted industrial end-users (in the residential sector as well as public and commercial buildings). NEEA is also responsible for the accreditation and development of the critical ESCO industry. NEEA faces a daunting challenge in scaling up its organizational capacity to undertake these responsibilities.

ESCO Industry Association

The South African Association of ESCOs (SAAE) was formed in 2003 to assist in the creation of a sound ESCO industry. The association has regular forum meetings at which matters of mutual interest are discussed. These meetings also promote a code of conduct to which members must adhere. All members of the SAAEs must be registered with Eskom DSM. There are currently 205 ESCOs registered with Eskom DSM of which only one is listed as inactive. A full list of registered ESCOs is available from the Eskom DSM website. This includes ESCOs and other entities such as redistributors. The SAAEs latest listing shows 12 associate members and 35 corporate members.

By March of 2007, these ESCOs had completed a total of 280 projects, all of which were subjected to a full measurement and verification campaign based on the International Performance Measurement and Verification Protocol (IPMVP).

A separate division in Eskom, Energy Audits, has been formed and tasked to independently evaluate, audit and report on the status of energy efficiency and load management activities. The measurement and verification of energy efficiency and load management savings, better known as M&V, activity is funded from the EEDSM funds collected through the electricity tariffs. The National Electricity Regulator (NER) which approves the budget requires an independent assessment and reporting process.

In order to increase the independence and credibility of the M&V

function, the site work is outsourced to South African universities. At the present, Energy Audits has contracts with seven universities including University of North West, University of Cape Town, University of Kwa-Zulu Natal, Tshwane University of Technology, University of Stellenbosch, University of Fort Hare, and Cape Peninsula University of Technology.

In the annual report from Energy Audits for the period 1 April 2006 to 31 March 2007, a total evening peak (18:00 to 20:00) impact of 423.9 MW has been achieved. A total of 864,577.5 MWh has been saved over the same period. This resulted in CO_2 emission savings of 864,577.5 kilotonnes annually.

Financing and Contracting

The guaranteed savings model is in use for all current EE and DSM-projects in South Africa with the investors being the national utility and end users.

Load shifting projects which impacted on the morning (7:00-10:00) and evening peak (18:00-20:00), with the focus on the evening peak, attracts 100 percent funding from Eskom. Energy efficiency projects are funded to the value of 50 percent.

There are three contractual arrangements in all EE and DSM projects. These can be summarised as follows:

The Eskom DSM contract with the client. This contract governs the relationship between the customer and Eskom DSM, and is the first contract in any DSM project that must be completed.

The Eskom NEC contract with the ESCO. Whereas the DSM contract is negotiated first, the NEC contract defines the relationship between the ESCO and Eskom DSM. It determines performance, payment and penalties.

The ESCO-client contract (maintenance). An important changeover point is the end of the performance assessment period. This is usually three months after construction completion, and it is during this time that the ESCO must prove that the project has met its contractual objectives. During this period the ESCO is liable for penalties due to under-performance. However, as soon as the project is performing, or over-performing, the liability shifts to the client to maintain the project at design performance.

Major Market Segments

The ESCO business in South Africa focuses on all the market segments; residential, commercial, and industrial.

The biggest impact in the industrial segment is made in the gold mining sector where underground lighting, water pumping, hot water systems, fridge plants, and rock winders are targeted.

The measure most frequently targeted in the commercial segment is efficient lighting.

ESCOs most frequently focus on hot water load control and efficient lighting in the residential market.

With Eskom embarking on Power Conservation Projects however, the focus in all the sectors will now be shifted. In the residential sector, the focus will be on rolling out programmes for efficient lighting, solar water heating, installation of aerated shower heads and geyser blankets, thereby reducing residential consumption of electricity. Efforts in the commercial sector will be concentrated on street lighting projects and the conversion of lighting, heating, ventilation and air conditioning systems. In industry, demand market participation contracts, process optimization and the promotion of the use of energy efficient electrical motors will be of primary concern.

Barriers for ESCOs

Cumbersome procedures for evaluating and approving project funding requests have delayed projects by months or even years. The EE/DSM Fund primarily relies on ESCOs to conduct EE activities, but there are still relatively few strong ESCOs. There is no mechanism for assisting the energy auditing and pre-feasibility study process, which means that ESCOs must assume considerable risk and time delays before projects can access financial support. As there is little or no commercial lending for energy efficiency, ESCOs have nowhere else to go. Some energy efficiency markets, such as small and medium enterprises (SMEs), and municipalities, just to mention a few, are virtually untapped.

There are some energy efficiency opportunities not covered by current implementation arrangements. Part of the problem is that the current DSM fund is not available for non-electric energy efficiency projects, even though more than two-thirds of final energy consumption is coal, oil and biomass.

Enabling Factors

Probably the most important positive enabling factor for the South African ESCO industry is that it has a strong backing from the national utility, Eskom. There are targets for demand reduction set by the DME for 2012 and 2025. Both these organizations realize that for these targets to be met, implementation arrangements need to be improved.

All ESCO projects done through the EE/DSM programme in South Africa are subjected to a full measurement and verification campaign based on the International Performance Measurement and Verification Protocol (IPMVP).

Future Expectations

In the beginning of 2008, major electricity cuts were experienced all over South Africa. This was a clear indication that the focus needs to be moved from load shifting projects to energy efficiency projects.

Since then the efficient use of electricity has become a national priority, a necessity for the future development of the South African economy and effective provision of electricity. Working towards these objectives is Eskom's Accelerated Energy Efficiency Plan that focuses on reducing electricity demand by 3,000 MW by 2012, and a further 5,000 MW by 2025.

In addition to the short term initiatives already discussed, there are medium and long term projects under consideration.

In the residential sector, applications for light emitting diode (LED) lighting will be examined. In addition, micro cogeneration, which allows consumers who generate heat or electricity for their own needs to move surplus power into the power grid, will be explored. The plan also provides for the examination of the use of generators, such as wind turbines, which are not now connected to the national grid.

In the commercial sector and the industrial sector the focus will be on led lighting. In addition, sustainable activities will form a significant part of the focus on medium and long-term planning.

Factors Affecting ESCO Development

The National Electricity Response Plan (NERP) includes a power conservation programme (PCP), which contemplates mandatory rationing. Preliminary socio-economic impact assessment has revealed that imposition of mandatory rationing will have a significant negative socio-economic impact.

Notwithstanding these findings there is common agreement that there is a need to maintain and improve on savings already achieved by voluntary means.

Eskom has done a significant amount of work with their key industrial customers in a range of sectors as to how increased savings could be achieved. Work done in the NERT economic impact task team has also identified a range of savings interventions which could be considered.

During an engagement between the Business Unit of South Africa (BUSA) and Eskom to explore alternative ways to achieve the savings goals of 10 percent, Eskom made a proposal for a voluntary rationing scheme. A minimum energy saving of 10 percent is set as the benchmark. Failure to meet the 10 percent target will have significant implications on the clients' electricity bill; e.g., a zero percent performance will result in an electricity bill almost double that compared to the baseline.

The major implication on potential bills, which could increase nearly 100 percent, is expected to cause industry to provide private funds to embark on energy efficiency projects.

Case Studies

A complete list of projects together with impacts made due to load shifting, energy efficiency, or peak clipping are available from the SAAE's website (www.esco.org.za). Project information includes ESCOs, clients, project type, location, intervention, and MW impact.

Chapter 5
The Middle East

Much of the world views the Middle East as an oil rich region and suspects it would have little interest in energy efficiency or in energy service company development. This is not the case.

The Middle East is not monolithic in its oil production profile. Countries of the eastern Mediterranean region, with the exception of Syria, produce and consume only modest quantities of energy. Israel, the Palestinian Authority, Jordan, and Lebanon import almost all of their petroleum requirements. Demand for natural gas is growing exponentially and electricity needs are also growing rapidly. Of great interest is the strategic location these countries occupy in terms of regional security and prospective energy transit.

In contrast, Egypt is a producer of both oil and natural gas with a decline in its oil reserves and a rapid growth in its gas reserves. Oil production continues to decline from its 1996 peak of 922,000 barrels per day of crude oil. As of 2004, almost all of Egypt's oil-dependent power plants had been converted to run on natural gas as their primary fuel. Now the sixth largest producer of liquefied natural gas (LNG), Egypt opened its first LNG export terminal in 2005. With the Suez Canal and Sumed Pipeline as strategic routes for Persian Gulf oil shipments, Egypt is an important transit corridor.

Scattered reports suggest most of the region is not particularly active in ESCO development. Some early discussion about energy efficiency and ESCOs have taken place in Saudi Arabia, but little has materialized. Mr. Walid Shahin reports that only one ESCO existed in Jordan, which focused on doing "shared energy auditing in industrial and commercial facilities. The ESCO stopped this line of work about a year ago because of financial problems/disputes with customers." As noted in Chapter 4, some activities have been reported in nearby Morocco and Tunisia, supported by the UNEP and UNDP.

We are fortunate to be able to present in this chapter quality reports from four countries; i.e., Lebanon, Israel, Turkey and Egypt, which represent very different ESCO development patterns in the Middle East.

Mr. Anwar Ali, Project Manager for the Lebanon Center for Energy Conservation Project (LCECP) with the assistance of Mr. Pierre El Khoury, the Energy Engineer for LCECP, have provided a clear overview of ESCO industry development in Lebanon where a project is funded by the Global Environmental Facility, managed by the United Nations Development Program (UNDP) and the Lebanese Ministry of Energy and Water, and executed by the Ministry of Energy and Water.

Mr. Z'ev Gross with the Israeli Ministry of National Infrastructure (MNI) has given us a different perspective regarding the ESCO industry. Mr. Gross has degrees in geology and law and claims that the secret of his accomplishments in the field of energy efficiency has been his ability to merge his earth science and legal backgrounds. He reports that MNI was mandated to deal with energy efficiency in Israel and chose performance contracting as the means of achieving its goals. With the consulting support of Econoler International, the government has been the driving force. Mr. Gross provides a fine summary of procedures governments can take to establish performance contracting and offers a current status report of ESCO development in Israel.

Mr. Larry Good, Dr. Halil Güven and Mr. Ata Osman Memik join forces to describe the evolution of the ESCO industry in Turkey. Their combined perspective offers a rich view of events leading up to the establishment of performance contracting in Turkey's unique situation.

In spite of help from donor agencies, Egypt has struggled for nearly a decade to create an ESCO industry. Mr. Emad Hassan, an Egyptian native, is a principal consultant with Nexant, Inc, a US-based energy consulting firm and currently heads up a global USAID energy efficiency initiative in five countries including Egypt. He offers a very succinct scenario of the barriers and struggles the industry has faced and ends by providing an optimistic prognosis for Egypt.

LEBANON

In 1992, after the end of 17 years of civil strife, the government of Lebanon launched an emergency rehabilitation and reconstruction plan for the country. The most prominent component of this plan has been the rehabilitation of the country's electricity sector, which to date has consumed 33% of the reconstruction budget. Investments have lead to an increase in power production capacity from 600 MW in 1992, to 1,400 MW in 1998 and up to 2,200 MW currently. However, with an estimated yearly growth rate of 4 to 6 percent in energy demand, interest in end-use energy efficiency and conservation is being seen as a viable alternative to meet the growth in demand. Accordingly, different private initiatives were launched in the late nineties to create energy service companies (ESCOs). Several companies were established to promote the ESCO concept. However, due to many constraints in the market as well as the unstable political situation, a large number of these companies have left the scene.

Until 2002, the only operational ESCO in Lebanon was National Energy Consultants (NEC). The company was founded in 1996 with a primary business focus on energy efficiency. Originally, the company started working closely in conjunction with Energy Automation Systems Incorporation (EASI) of the United States of America. This affiliation served as a solid booster towards offering turnkey energy efficiency solutions to the Lebanese market.

The Lebanese Center for Energy Conservation (LCEC)

In 2002, the project "Lebanon Cross Sectoral Energy Efficiency and Removal of Barriers to ESCO Operation" was initiated. The project is funded by the Global Environment Facility (GEF), managed by the United Nations Development Program (UNDP), and the Ministry of Energy & Water (MEW).

The goal of this project is to reduce green house gas emissions in Lebanon by improving demand side energy efficiency through the creation of a multi-purpose Lebanese Center for Energy Conservation (LCEC). The Center intends to simultaneously undertake barrier removal activities and provide energy efficiency services to the public and private sector industries to assist these firms in becoming independent, commercially viable private corporations. There will be a broad range of supporting activities including information dissemination, awareness programs, policy analysis, and program design.

A major component of the LCEC activities consists of conducting and financing a number of energy audit studies with identified and targeted end-users and end-use technologies/applications. The major objective is to create a new market for energy audit applications, thus offering an opportunity for a number of engineering firms to work as ESCOs.

The ESCO Business in Lebanon

Among the major activities of LCEC, energy audits constitute an essential part. A prequalification process for energy audit firms in Lebanon was launched in March 2005 to identify potential energy audit/energy service companies capable of performing energy audits to meet LCEC standards and professional requirements. A prequalification questionnaire was developed and disseminated to collect adequate information on local Lebanese companies with regard to their capability and experience. The information was used by LCEC to develop the list of candidates for energy audit contracts. This prequalification process is a continuing effort to identify all potential ESCOs in Lebanon.

Training Workshop for Energy Engineers in Beirut

Currently, a number of selected engineering firms are working with LCEC, conducting energy audit studies. Currently, only one company has the ability to finance the proposed energy conservation measures (ECMs). National Energy Consultants (NEC) finances ECM implementation from their own resources and loans from local banks. The other companies, Apave Liban, Beta Engineering, Energy Efficiency Group (EEG), Metacs, SGS Liban, and Tecmo are engineering firms working on developing a department for energy services. It is important to note that LCEC is working on the creation of a national fund to be used for the financing of typical

energy saving projects.

The Lebanese market for energy saving and conservation measures is, therefore, centered mainly on the activities of LCEC. Although LCEC is trying to expand the market and create new opportunities, it is still too early to say that a market for ESCOs exists in Lebanon.

Mostly, the work of the ESCOs is focused on finding solutions for major industries or commercial facilities. In that respect, solutions target lighting systems, motors, HVAC systems, automation systems, building management systems (BMS), building envelope, pumps, machines, power factor correction, and others.

In the near future, it is expected that the ESCO business will be developed due to the rapid increase in the oil prices. In addition, there is a real commitment from the Government of Lebanon to adopt energy efficiency and renewable energy solutions as part of its energy reform. It is important to note that the government of Lebanon still subsidizes electricity tariffs in the country.

ISRAEL

The Israeli ESCO industry is, at the time this is written, at a crossroads. The industry, along with the performance contracting methodology, which serves as its primary modus operandi, is at the heart of the most ambitious energy efficiency effort ever carried out in Israel. Israel's government has resolved to attain an annual saving of 16 TWh (1 TWh = 1×10^9 kWh) in the year 2020, with an accrued saving of 104 TWh by the end of that year—two and one half times the overall consumption of the State of Israel in its base year—2006.

The effort to introduce performance contracting and support a nascent ESCO industry was actually the result of a unique technicality of Israeli law. Under Israeli law, legislation must state explicitly that it applies to the state in order to apply to government. The Israeli legislation regarding energy efficiency does not include this clause, and, therefore, attempts to apply regulations aimed at energy efficiency to government facilities met with resistance from the Ministry of Finance (MOF).

The Ministry of National Infrastructures (MNI), the ministry with the mandate to deal with energy efficiency, decided, therefore, to introduce energy efficiency via performance contracting methodology, which, MNI felt, could be introduced on a voluntary basis.

The ESCO concept was known in Israel, only in a very superficial way, and the risk balance aspect of the methodology, as well as the understanding of what might act as barrier to the methodology, was not understood. As these were never addressed, the methodology simply never developed roots.

As a first step, MNI retained the services of Econoler International as consultant to survey Israel's potential ESCO industry and the potential market. The firm's conclusion was that the technical expertise was available in Israel to support an ESCO industry and that the relevant capacities would have to be developed in order for the methodology to be successfully implemented.

MNI held a number of workshops aimed at various actors—essentially the potential ESCOs and certain clients. The financial sector was not approached at the outset.

MNI then initiated a pilot project. Upon the advice of its consultants, the pilot was carried out using the guaranteed savings mechanism, and financed from MNI's budget. Four facilities were chosen, representing three different ministries, essentially on the basis of the relative age and state of their infrastructure—two major police facilities, a facility of the Taxation Authority and a building in one government hospital. It is interesting (but not surprising) to note that entities with an entrenched incentive plan for energy efficiency (the Taxation Authority and the Police) were immediately receptive to the idea of hosting the pilot, whereas the entity that did not have such a plan (the hospital), was not forthcoming to the proposal, and was in constant conflict with the ESCO.

The pilot projects—which were small, due to the very limited budget—did manage to reveal a number of barriers facing the ESCO driven, performance contracting methodology. (i) The difficulties posed by the government purchase/tender procedure proved to be a major, costly barrier. The tender system essentially mandated multiple IGAs (Investment Grade Audits), which rendered great risk to the nascent ESCOs. As a result, the project was required to set aside funds in order to also pay ESCOs that were not awarded a project. (ii) The ESCOs were not completely prepared for the performance contracting methodology. As a result, the IGA-driven projects were not, in certain instances, sufficiently well designed for smooth implementation.

Other barriers became apparent also. In discussions MNI had with potential financiers, as well as within the framework of a workshop given to the financial community, it became clear that the performance contracting methodology was unknown to them. In addition, projected transaction scopes were insufficient to warrant what these institutions viewed as the requisite due diligence process. All this, combined with the relative inexperience of the ESCOs and their thin portfolios, would make the financing costs to the ESCOs prohibitive. It was clear that government intervention was required.

MNI then commenced proposing what it labeled the "economic environment" required to mitigate the perceived risks in the energy efficiency performance contracting market. MNI proposed a set of "tools" that would act to mitigate these risks. These tools included tax incentives, revolving loan and guarantee funds, technical assistance and regulatory tools. These tools were presented in documents prepared both by Econoler International and MNI and presented to MOF, as well as to other stakeholders. Initially, these tools were received quite apathetically and the fear was that the effort would simply evaporate.

What made the difference, ultimately, was the pilot projects. The most important result of the pilot projects was the visibility given the performance contracting methodology vis-à-vis Ministry of Finance (MOF). The relevant departments of the MOF—especially that of the accountant general—had always been party to presentations on the performance contracting methodology, on ESCOs and on the importance of government intervention in the area of energy efficiency. MOF representatives were also always invited to take a role in the capacity building initiatives taken by MNI. With the initiation of the pilot projects, MOF's interest grew even further.

In light of all the above, Ministry of National Infrastructure proposed that the initial performance contracting effort take place in the government sector. This would have a number of advantages: (i) The "client" risk perceived by the financing community; i.e., the risk presented by the ESCOs' clients, would be mitigated (as the government is a "good" client); (ii) Any investment made by the government would be aimed first at cutting back its own energy expenses; (iii) the initial round of government intervention would only have to be aimed at the ESCOs, rather than both at the ESCOs and the private sector clients. This proposal was favorably advanced when two government entities, the Police and the Ministry of Health, turned to MNI with requests for technical assistance in carrying out relatively large-scale, shared savings performance contracting, projects.

At the time this is being written, the situation in Israel with regard to ESCOs and the performance contracting methodology, is as follows:

- MNI has issued a call for ESCOs to identify themselves and to be registered. Registration criteria has not been particularly stringent. All ESCOs were required to sign a code of ethics which was modeled after the New Orleans Municipality Code of Ethics. Eighteen ESCOs registered in response to the call, which remains open to this day.

- MNI has incorporated performance contracting into its "Assistance to Micro-projects" program, which gives grants to various energy efficiency projects. The mere undertaking of an energy efficiency project using the performance contracting methodology is sufficient to grant prima facie eligibility for an award. In 2007, approximately 35 percent of all projects awarded grants under the program were performance contracting (shared savings) projects.

- The tools proposed by MNI for the mitigation of risk in the ESCO business space will be discussed and hopefully approved by the government within the year.

- Two government entities with relatively many and large facilities, are now in contact with MNI for technical assistance in carrying out projects in their facilities. It is hoped that the tenders for these projects will be issued soon.

- MNI is seeking to further enhance the capacity of all stakeholders in this market. It will play a major role in assisting any government/public entity in the development of energy efficiency activity, and will also spearhead the broadening of risk mitigation tools into the private sector.

- MNI will seek to leverage the capacity and expertise gained from its efforts in this area into a tool for regional cooperation and development.

TURKEY

Mr. Larry Good, Dr. Halil Güven, and Mr. Ata Osman Memik of Envo Energy Services all contributed to the following valuable insights on the ESCO industry in Turkey.

Background

Turkey, which has a young and growing population, low per capita electricity consumption, rapid urbanization and strong economic growth, has been one of the fastest growing power markets in the world for almost three decades. Projections indicate that rapid (approximately 8 percent annually) growth in electricity consumption will continue until 2015. Turkey is a member of the International Energy Agency (IEA) and has also signed the European Energy Agreement and Energy Efficiency Protocol. Turkey is in the process of signing and ratifying the Kyoto Protocol.

Starting in mid-1980s, a US-based energy company, under a contract with the Turkish Electricity Authority, mapped the renewable energy power generation potential of Turkey. The company also introduced the concept of industrial energy audits to Turkey by performing a series of audits in several industrial facilities throughout the country. This was the first attempt at launching an ESCO in Turkey. Subsequently, these and other developments led to the recognition of the importance of energy efficiency, which resulted in the establishment of an "Energy Conservation Measures (ECM)" division in the Turkish Directorate of Electrical Power Resources Survey and Development Administration (EIE), and a National Energy Conservation Center (NEEC) in the Ministry of Energy and Natural Resources. However, the early ESCO attempt failed due to lack of appropriate support mechanisms in the country.

In the 1990s, the EIE started work, with a grant from Japan International Cooperation Agency (JICA), to develop a 10-day training course on management of energy efficiency in industrial facilities and buildings. With limited resources, NEEC tried to increase the public awareness. After 1995, EIE started training hundreds of engineers as energy managers. However, the need was too great to be met with these limited efforts, and there were still no laws and regulations to encourage conservation or support ESCOs. In 1997, the International Energy Agency (IEA), in its Turkey Review Report, recommended to the Government of Turkey to implement the following measures:

- Take active steps to better account for the energy savings in investment decisions and to better coordinate policies within the administration;
- Continue to provide public information about the benefits of increased energy efficiency and about the different measures to achieve these improvements.;
- Take active steps to involve all large industries in the energy efficiency program and carefully assess the results of the program. In addition, encourage smaller enterprises to save energy; and

- Simplify administrative procedures to better allow the development of cogeneration and consider encouraging district heating.

Subsequently, the Turkish government issued a decree requiring industrial facilities with a total annual energy consumption of 2,000 TEP (tons of equivalent petroleum) or more to either establish an in-house "energy management unit" or to employ an "energy manager." Due to lack of regulations and incentives, as well as energy efficiency and auditing services from third parties, ESCOs, were neither viable nor available. During this period, USAID also tried to assist Turkey in developing the appropriate laws and regulations, and to provide training support to launch industrial energy audits and ESCOs.

The very first ESCO in Turkey, National Britannia Energy Management Services, was established by a UK-based firm in 1999. The parent company, National Britannia, had a strong reputation for its Environmental Hygiene and Safety consultancy services in the food and health sectors as well as hotels and shopping malls. As large consumers of electricity and heat, these were potential clients for energy efficiency and/or ESCO projects. This attempt lasted almost three years and resulted in a number of energy audits performed in several medium- to large-sized factories, hospitals, hotels, and shopping malls. National Britannia implemented some projects in a shopping mall in Istanbul. The activity ceased in 2002 after repeated financial crises in Turkey and a lack of energy efficiency laws and regulations to support ESCO activities.

In 2005 the IEA wrote: "ESCOs are not active in Turkey. The main problems appear to be a lack of appropriate regulation and Turkey's high inflation." Since then ESCO activity in Turkey has restarted. In 2006 GlobalNet Energy Services International (GESI), a branch of the American company GlobalNet Venture Partners, LLC, began probing the market with industrial energy audits. In 2007 Envo Energy Services, a branch of the Turkish company Envo Group, Ltd. (affiliated with National Britannia-Turkey), took over the GESI operation and expanded both the range of services and the market sectors. The certified professionals to introduce the first ESCO services into Turkey came from America.

The lack of regulation mentioned by IEA is currently being rectified. The Government of Turkey recently passed an energy efficiency law, which is expected to provide the framework for ESCOs, and its regulations are expected to be finalized this year. When finished, the regulations will define qualification criteria for accrediting ESCOs and certifying energy managers/auditors. Most of the support and technical input for this development came from EIE and JICA. This new law will also decrease the 2,000 TEP threshold (mentioned above) down to 1,000 TEP.

ESCO Industry and Market

The new law sets a target of 15 percent savings in Turkey's overall energy consumption by 2020. With current prices, the annual savings potential is approximately USD 4 billion. Initially, ESCOs may be able to generate efficiency improvement projects and energy performance contracts to achieve 10 percent of the potential savings. The size of the market at this level will be USD 400 million per year. In full swing, with a fully developed ESCO industry, the market size could be much higher.

Since regulations are still under development at the time of this writing, a list of qualified and licensed ESCOs does not yet exist. Of course, many companies sell and install equipment. However, the Envo/GESI initiative is possibly the first commercial effort in Turkey to focus exclusively on the classic energy services concerning energy efficiency and renewable energy in this decade.

There are also many companies selling energy efficiency equipment and products, such as economizers, variable frequency drives, solar thermal, LED lights, combustion chillers, steam traps, analyzers, and VRV systems. But none of these companies are able to do an investment grade energy audit and do not function as ESCOs. They are mainly agent/distributor of their product and give partial energy auditing related to their areas.

Financing and Contracts

Today the only available contracts are for energy efficiency consultancy, equipment installation, and energy production and distribution. The shared and/or guaranteed savings contracts are expected after the release of the energy efficiency regulations. Funding for consulting contracts comes from the clients. For energy production there are available bank loans and leasing of equipment. Especially in renewable energy production, the government gives a 10-year guarantee for purchasing supplied power at the average wholesale selling price of the previous year.

Types of Projects

Currently, there is no clear definition of sector activities. The majority of ESCO projects are not based on a full, investment grade energy audit, so they only cover partial energy systems; e.g., steam systems, lighting systems, industrial and building automation, waste heat recovery, boiler efficiency, motors, HVAC, renewable energy (solar & wind), cogeneration and other energy systems. Most companies ("ESCO candidates") do partial energy studies and in most cases represent a brand of equipment.

Major Market Segments

Due to the regulatory void, currently there is no organized market activity. The ESCO industry is expected to pick up once the by-laws are released and the energy law is in full implementation.

So far, the most receptive sectors have been resort hotels of Turkish ownership and multinational companies in other sectors. Resort hotels on the Mediterranean and Aegean Seas generally do not have access to inexpensive natural gas; so as tariffs rise, hoteliers are paying more attention to their energy expenses.

International manufacturing enterprises (in tobacco and textiles, so far), as well as international retailers, have shown a more global vision of environmental issues than Turkish-owned companies.

Barriers

Even without incentives, there are plenty of energy saving opportunities which are already feasible on a purely commercial basis, but Turkish companies are not aware of how to achieve the benefits. The barrier is fear of high initial investment without consideration of the return. An internal rate of return (IRR) of 30 percent or more, quite common for energy saving measures, should be attractive to any business. However, technical personnel do not know how to translate energy savings potential into financial feasibility to gain the confidence of investors. ESCOs can help here as quickly as Turkish clients understand and accept the help. Also, increasing tariffs will drive change faster.

Enabling Factors

Demand side management (DSM) incentives, which had been lacking in Turkey, are coming soon from the new energy law. The impending regulations will spell out incentives, especially buying down the cost of energy studies. This will help move the ESCO market forward.

The new Energy Efficiency Law should be the biggest enabling factor for the ESCO industry. The new legislation provides a framework for the certification of the ESCO industry and the guidelines for the financing, as well as incentives (subsidies), for energy projects and EIPs. Currently, the industry is waiting for the regulations to come out.

Project development and implementation is, of course, the usual business of ESCOs. In an emerging market, without the appropriate legislation in place, however, project-based business can mean feast or famine.

In addition to achieving more, staff development programs are simply a good marketing tool for projects. Clients are more receptive to ESCO recommendations after training and team building teach managers how things work and what to expect. Projects will get started that otherwise would have been blocked by fear or lack of understanding. Staff development is a less expensive service than project development and management, but it helps an ESCO make payroll and keep operating while waiting for projects to come in the door.

Future Expectations

In the short run, Turkey is focusing on energy savings. If Turkey can launch an effective ESCO industry, savings of 8-9 percent per year could be found in Turkey's annual overall energy consumption, which could defer capacity needs. In addition to lack of awareness by the public at large and a lack of effective industrial energy efficiency measures, currently Turkey looses close to 20 percent in power transmission. Therefore, there are plenty of opportunities for ESCOs in Turkey. It is very likely that Turkey can "generate energy" from 8-9 percent energy savings for a number of years to meet the needs of its economic growth.

EGYPT

History

In Egypt, the ESCO industry has been on a slow growth pattern since the late 90s when it was initially introduced through various donor-funded projects, namely by the USAID and UNDP. Much of the awareness and capacity building activities offered led to the successful formation of a non-governmental organization (NGO) to represent the industry and help expand EE market opportunities.

The Egyptian Energy Service Business Association (EESBA) is a registered NGO established in 1999 to promote the interests of companies working in the Egyptian energy sector and to support the drive for greater energy efficiency. EESBA's membership includes about a dozen companies which provide energy products and services to the market, including equipment suppliers and consulting firms. The association actively participates in policy development, links key business groups and government leaders, encourages EE partnership initiatives and helps build capacities to meet market needs. EESBA has cooperated with the Association of Energy Engineers to develop a local version of the Certified Energy Manager (CEM) course for local delivery by Egyptian energy engineers. Approximately 300 engineers have been certified by AEE as CEMs.

Current ESCO Industry Profile

A market assessment was conducted recently by Nexant, under contract to USAID, to review the availability of energy efficiency service providers and their capabilities to deliver energy service solutions. A total of 35 entities were identified as providers of energy services including ESCOs, equipment suppliers, electrical/mechanical contractors, engineering consulting firms, and other relevant professionals providing demand-side services. In-person interviews were conducted with 19 companies that were characterized as 'large' by local market definition, while feedback from market experts served as the source on the capacity of the remaining small to mid-size companies. The assessment focused on 1) the most commonly used energy efficiency applications (lighting, power factor, water heating, automated controls and HVAC), and 2) on the key elements of the ESCO offerings such as energy audits, equipment supply, project management/implementation, M&V, commissioning, financing, and performance guarantee.

Large Energy Service Providers

This group of 19 companies provides various aspects of energy services; however, none are offered under a performance-based arrangement. A little over 50 percent of these companies have the capabilities to provide services under most of the commonly known EE applications, 25 percent are capable of providing limited EE services, and the remaining 25 percent could provide individual services; i.e., implementation or consultation only, or one end-use technology. 20 percent of this segment indicated unwillingness to provide project finance. In general, experience in risk-based services and M&V protocols was limited, and most of the contracting capabilities seemed to be geared towards provision of specific equipment wrapped with a product guarantee.

Small Energy Service Providers

Very few small and medium-sized companies have experience with the full range of energy efficiency products and services, and they typically shy away from performance-based arrangements. They have significant challenges in obtaining project finance except for a small number of transactions. However, they usually serve as a convenient resource as subcontractors of specific products or services for the larger group of providers.

The current capabilities of the energy service sector indicate that implementing a true performance-based contracting under current market conditions is still a challenge. However, most, but not all, of the ingredients necessary for this type of delivery approach exist within a small number of service providers. Further, testing of different business models, which could push the market toward a true performance-based delivery is still recommended.

Egyptian and other multinational service providers have a high level of comfort with the supply of several energy efficiency products as well as the skills necessary to estimate technical performance. Assumption of financial and technical risks, however, appear to be undesirable by providers as the market is still unwilling to pay the associated premium.

Performance-based contracting has been used in a very limited number of cases since 1996, but without a real success story to document. Consequently, knowledge of measurement and verification techniques is limited, and even those, who are familiar with it, are reluctant to promote it and fear the lack of understanding on the customer's end. There are no current protocols being used, but the UNDP-sponsored energy efficiency project has produced a simplified version of the international M&V protocol. Since its inception in 1999, the UNDP project provided significant support to the struggling ESCO industry including partial incentives for conducting audits, support to credit guarantee mechanism to support project finance, and support in evaluating project and product performance.

Energy Efficiency Potential in Egypt

It is rather difficult to assess the real size of today's products and service market for energy efficiency in Egypt, especially with the recent government move to gradually reduce the end-use energy subsidy. The only thing that one can be certain of is that the demand on EE will grow substantially with the increase in energy prices. For years, the energy subsidy provided a disincentive to end users to reduce or rationalize their energy consumption. A study conducted in 1998 by Bechtel Consulting (currently Nexant) under contract with USAID, estimated the potential investment in EE to exceed USD 1 billion. Only a fraction of that potential has been realized due to the subsidy disincentive.

An analysis of energy and economic indicators prepared in 2000 revealed that, when compared to some similar economies, Egypt's progress towards a more energy-efficient economy needs help. Over the period 1998 to 2003, both Egypt's GDP and the primary energy consumption rose at approximately the same rate, indicating no changes in the energy intensity. This clearly shows that efficiency improvements were not enough to affect the country's economic development.

Energy efficiency is likely to gain more attention in light of the various market drivers currently unfolding. First, Egypt is currently privatizing many of its state-owned industrial enterprises, leading investors to rationalize energy consumption in order to remain competitive. Second, the Government of Egypt has also added EE as an integral part of its future supply/demand energy balance through 2020. Some 8-10 percent of the growth in energy consumption by then is expected to be met through improved efficiency. Finally, the gradual increase in energy end-use prices will put efficiency in a sharper focus.

Various donor-funded reports have attempted to estimate the potential for energy savings and the associated investment requirements. While numbers varied considerably, the most rational conclusion of these reports placed the potential in the range of 20-40 percent of the annual primary energy consumption.

Financing and Credit Guarantee

Financing energy efficiency (EE) projects remains one of the key challenges facing the ESCO industry in Egypt, mainly due to the inherent incompatibility of the existing lending vehicles with the nature of EE projects. While banks are eagerly looking for investment opportunities, EE projects have not yet received their desired attention. Banks and leasing companies are more interested in low-risk clients with good potential for profit, and are reluctant to invest in project finance for the small and medium-sized enterprises. Additionally, there is less enthusiasm about using the energy measures as collateral as they become part of the real estate property once installed. The economic benefits realized from these projects are also considered reductions in operational costs and not additional generated revenue that financial institutions would be interested in to secure their investments.

Many donor-supported activities have attempted to entice the financial industry to bring investments into EE projects, which can leverage the knowledge and capabilities of ESCOs, but few are being translated into experiments.

UNDP's Credit Guarantee

The UNDP-sponsored energy efficiency project (EEIGGR) provided a financial guarantee to one of the local credit guarantee companies to enable ESCOs access to commercial lending for EE projects up to USD 180,000. The guarantee is offered by the local Credit Guarantee Company (CGC), which was established in the early 90s by a group of Egyptian banks to support small and medium-sized companies in various sectors. CGC provides back up guarantees to banks on behalf of a network of pre-qualified ESCOs to receive a 70 percent guarantee for their requested loans. The CGC guarantee also provides a safety net against non-payment from the project owner/host. Several projects have already started to benefit from this mechanism, which is expected to make bankers more comfortable with EE projects.

Development Credit Authority

The Development Credit Authority (DCA) is a mechanism established by US Agency for International Development (USAID) to provide a tool for their missions around the world to encourage the use of credit and expand financial services in underserved markets. DCA helps to mobilize private capital for financing development initiatives. Using the DCA tools, such as loan guarantees and loan portfolio guarantees, USAID missions are able to partner with lending institutions in making resources available for investments that support development objectives. This, in turn, demonstrates the economic viability of such investments to the local banking sector creating sustainability in the long term.

In 2004, USAID/Egypt signed a $10 million loan guarantee agreement with the National Société Générale Bank (NSGB) to support the development of energy efficiency and eco-tourism projects in Egypt. The agreement provided NSGB with a guarantee of up to 50 percent on borrower repayment of loans for eligible projects to encourage the bank in offering favorable financing to the market.

Types of ESCO Projects

With few exceptions, most energy efficiency activities are provided as part of other electrical and mechanical retrofit services when equipment nears the end of useful life. However, high electric rates for commercial establishments, rate penalties for low power factor, and price differential between fuel oil and natural gas created strong incentives for energy end users to implement lighting efficiency projects, fuel switching, and power factor improvement as stand-alone projects. While ESCOs do implement projects in these three areas, most do it under unit rates, and some product warranty, but never with performance-based rewards and penalties.

Many commercial and industrial energy audits have been performed over the past 20 years, but only a limited number have made it to actual implementation. The energy engineering skills are quite strong in Egypt, but subsidized energy pricing has been a major disincentive for efficiency. Only in specific sectors such as commercial establishments (hotels, retail stores, office buildings, entertainment facilities, etc), where the tier pricing structure helped raise the energy cost to end users, the need for conservation solutions has been high.

Distribution companies are now implementing projects using different versions of the utility/ESCO approach. Some electric distribution companies and natural gas distribution companies offer clients, in their geographic areas, integrated services in the three main areas mentioned above, and charge them on the utility bill over a period of time. This model provides less risk and more profit for the distribution company, especially in the natural gas area, as they profit from the margin on the project implementation as well as the revenue from gas sales.

Investment requirements for EE projects in Egypt vary considerably. The UNDP's EEIGGR project helped support approximately 200 audits of which approximately 60 percent needed over $180,000 to implement, 17 percent required under $18,000, and the remaining projects were in between.

Major Market Segments

The large purchasing requirements and generally low credit risk of government agencies make this sector an attractive market segment for energy service providers. The public sector in Egypt consumes approximately 5.5 percent of total energy use with approximately $180 million in lighting consumption. However, it is one of the most difficult sectors to penetrate given their rigid public procurement regulations and the absence of profit/loss culture for facility management professionals. A current pilot project by USAID in cooperation with the UNDP will test a replicable model for procuring energy efficiency measures in public buildings through energy service companies to reduce the investment burden on the government budget.

The main idea of the pilot is to procure energy efficiency services from existing providers through a competitive bidding process that ties compensation to the performance of the installed measures. The success of this pilot is expected to 1) encourage government agencies to use this procurement model thus reducing their energy consumption while improving their operational conditions, 2) identify changes to the government procurement regulations that could help maximize the use of this performance-based procurement, and 3) to open a large market for energy efficiency products and services that would increase availability and reduce costs of such services.

The pilot succeeded in including an M&V plan and attached it to the ESCO final payment. If replicated as a model for all public facilities, an investment of approximately $100 million will be needed just for lighting efficiency measures.

Market Barriers for ESCOs

The key barriers for market penetration of ESCOs include:
- Subsidized energy prices that remove the incentives for end users to consider implementing EE projects (although this will be phased out);
- Lack of incentives to adopt energy efficiency technologies in the public sector, which is the most attractive segment;
- Government procurement rules are product-focused and do not encourage integrated energy savings and or performance-based approaches;
- Lack of compatible project finance mechanisms; and
- Limited awareness of the value that ESCOs bring. End users and project owners typically determine the feasibility of projects on first-cost and not on a life cycle basis.

Promising Factors for a Growing EE Market

In Egypt there are a number of major factors affecting the potential EE markets:

- Gradual reduction of subsidy;
- In-progress, new electricity law allowing private generation facilities and bilateral contracts between end-users;
- Interest in developing and enforcing EE building standards;
- Abundance of available capital for investment;
- Advanced engineering skills;
- Rising interest of the Government of Egypt in EE to close the supply/demand gap; and
- Privatization of state-owned energy-inefficient enterprises.

Future Outlook

The demand for energy efficiency in Egypt is likely to increase at a faster pace than the ability of existing energy service market can handle. Increasing energy prices across all fuels, especially in the energy intensive industries, coupled with the ongoing privatization process of state-owned industrial enterprises, are strong drivers towards efficiency. With government future plans targeting a GDP growth of 7 percent per year, energy consumption could double every 10 years. Competition by suppliers of energy resources by existing and new investors will become more challenging and will require efficiency solutions to reduce costs. The government has accepted energy efficiency and renewable energy as main contributors to its supply/demand energy balance through 2020.

Demand for energy services will continue in the current active areas of lighting and other known measures. However, demand for services and solutions is likely to grow fast in the natural gas and renewable energy areas. Shifting energy reliance from liquid fuel to natural gas will require an infrastructure of services to support all related market demands. Natural gas local distribution companies are struggling to expand the gas network fast enough to meet demand across the country. This will certainly require more customized solutions to end-users for their specific use.

The other area for future growth is renewable energy. Solar, wind and biofuels have strong potential in Egypt but the service market is not well established to support growth in this sector. Therefore, ESCOs with expertise in these areas and with an appetite for green field operations will certainly be in high demand.

We anticipate that financial institutions will design more specialized lending mechanisms to support energy efficiency projects as the market demand increases.

CASE STUDY

USAID and UNDP supported the implementation of a pilot project to promote EE procurement in public sector buildings. The project was hosted at a 22-story headquarters office complex of one of the government ministries located in Cairo. The pilot objective was to test a replicable model to enhance existing government procurement to include EE measures and to leverage the ESCO approach to reduce the burden on the government budget. To focus the process on the procurement practice, the project was limited to simple and proven lighting efficiency measures, such as fluorescent lamps, electronic ballasts and CFLs. A preliminary energy audit was conducted, and was followed by an RFP, to which 3 ESCOs submitted proposals. An ESCO was selected for a project cost of $67,000. The audit showed an annual consumption of approximately 2.5 million kWh and a peak demand of 1,200 kW with lighting responsible for 53 percent of total consumption.

Typical government RFPs include specific count of energy measures, locations, minimum product specifications and rigid payment stream that is focused on the delivery of the measures and not on the associated results. Lengthy negotiations with the procurement officials and facility managers for the project led to the addition of a few performance-based clauses that tie a portion of the project cost to an M&V plan for several months after project implementation. A few after-installation measurement reports indicated savings of approximately 20 percent of the total bill for the complex, or some 45 percent reduction in lighting consumption. Lighting quality was improved significantly and was documented in the M&V reports. However, certain installed measures resulted in a higher power factor readings than the allowable standards, which resulted in withholding the final payment until these measures were replaced.

The main achievement of this pilot was to convince public agencies that a semi-performance approach can be integrated into public procurements and produce not only operational savings, but also protection against undesirable performance.

Engineering services are quite advanced and inexpensive, thus offering an advantage for ESCOs for audits, feasibility studies, and M&V services. Performance contracts as they are applied in Europe and the U.S. can be risky for ESCOs to use, as contract enforcement is not an easy task. The credit guarantee mechanisms can help the ESCO market in overcoming the financing and credit risk issues.

In general, most commonly known energy efficiency technologies and related services are available in Egypt, but the main barrier to expand performance-based type services remains to be the limited availability of compatible long-term project finance options and the limited experience with performance and risk-based delivery approaches.

Chapter 6
Asia

Asia is becoming a voracious consumer of energy. While portions of Asia, such as China, have enormous potential to provide energy domestically, the region is a very large energy importer.

Asia is expected to become even more critical to the world's energy outlook in the years ahead. By 2030, the world consumption of energy is expected to increase by 50 percent. Of that increase, India and China are expected to contribute 45 percent of that consumption growth.

As price tensions increase, the window of opportunity for energy efficiency, and often ESCO growth increases as well. At present, China's energy utilization efficiency is only 33.4 percent, nearly 10 percent lower than the advanced international level. From the reports, it is clear that for both economic and environmental reasons, the countries in Asia must continue to aggressively seek ways to increase energy efficiency and reliance on renewable energy.

Asian countries vary tremendously in the degree of ESCO industry development. The reports presented in this chapter illustrate the great contrasts from the methodically state-supported growth in Japan to no significant action in the Philippines.

We are very fortunate to tap into the rich data sources the Japanese Association of Energy Service companies (JAESCO) have, through the report authored by Mr. Chiharu Murakoshi, Mr. Hidetoshi Nakagami and Mr. Takeshi Masuda. The sequential scenario of government actions, including those addressing environmental concerns, provides very useful information as to the steps others might find helpful in securing government support for ESCO development. The description of JAESCO's activities also offers very helpful guidance to new associations, or those that wish to serve their members more effectively.

This chapter also offers a broad spectrum of ESCO activities in smaller countries including Thailand, Philippines, Vietnam and the Republic of Korea (South Korea). Mr. Arthit Vechakij, managing director and Mr. Raumlarp Anantasanta, deputy managing director of Excellent Energy

International Company, Ltd., have provided wonderful insights into the ESCO situation in Thailand. Dr. Alice B. Herrera has described the ESCO status in the Philippines while Mr. Ed Sugay of Siemens offers us a short overview of ESCO activities in Korea and Mr. Ha Dang Son gives us insights as to what is transpiring in Vietnam. While this certainly does not cover all the ESCO activities in Asia, it offers a very good representation of the range of ESCO activities in the smaller Asian countries.

A summary of activities in India's public sector has been drawn from a recent report by Econoler for a World Bank funded IREDA program with valuable input from Mr. D.K. Satya Kumar. Ms. Nisha Menon offers a brief view of what her company, DSCL Energy Services, has accomplished in the private sector.

The chapter concludes with a very informative look at ESCO progress in the Peoples Republic of China. Mr. Shen Longhai, executive director, and Ms. Lily Zhao, secretary general, of EMCA, the Chinese ESCO association, offer an assessment of the ESCO industry, the barriers currently faced and the opportunities ahead. Mr. Shen and Ms. Zhao describe the growth from a nucleus of three energy management companies to the large and growing industry of today. A scenario is offered from the early World Bank support, through training and consultation provided by Kiona International (now Hansen Associates), to the three original ESCOs, to the training they, in turn, have given the new companies joining the industry today.

The countries represented in this chapter have a total population of nearly three billion; so it is not surprising that the concerns and opportunities afforded local and international ESCOs in this market constitute a major chapter in this book.

JAPAN

The energy service company industry is expanding in Japan, started by a feasibility study initiated in 1996. From the beginning, the industry was intended to play a role among various governmental measures against the global warming, and has been developed as a type of private industry supported by the government.

The energy saving potential of existing buildings is considered to be enormous. Although energy saving retrofit work has been practiced, so far it has not necessarily functioned effectively. The ESCO industry is expected to become *the* effective tool, in addition to governmental regulations, to develop an energy saving market with huge potential.

It is often difficult to advance the argument for energy saving retrofits due to the long payback period and the perceived lack of investment recovery, unless documented energy savings are realized. Since energy saving retrofit varies in each existing facility, and energy efficiency measures adopted to save energy are also diversified, the owners of facilities are confronted with difficulty in funding effective solutions, and in making investment decisions. Moreover, it is exclusively the owner's risk whether the saving of energy is attained and expected investment recovery is realized.

Under the energy saving guarantee mechanism offered by the ESCOs, owners of facilities can recognize the prospect of investment recovery even without profound knowledge of technical matters, and are provided with significant information by the ESCO so as to be able to make investment decision appropriately. Among various characteristics of an ESCO, guaranteeing the amount of energy savings is the most important, and has become a critical differentiator between ESCOs and ordinary retrofit projects.

The industrial sector and the commercial sector are currently the targets for ESCOs in Japan. Although the residential sector has been examined to introduce ESCO projects, it has not been realized yet.

The Introduction of ESCOs in Japan

In the 1990s, while the ESCO model progressed in the US and was reintroduced into Europe, the Advisory Committee on ESCO Investigation in Japan was established within the Agency of Natural Resources and Energy under the Ministry of Economy, Trade and Industry. Various programs were carried out, such as feasibility studies, experimental projects, examination of the measurement & verification protocol, ESCO's introduction to local authorities, and establishment of the Japan Association of Energy Saving Companies (JAESCO). The implementation of these programs has helped the ESCO industry expand.

ESCO Introductory Study (1996)

An advisory committee on ESCO investigation was established in the Agency of Natural Resources and Energy (METI) in 1996. This examination was designed to introduce ESCOs to Japan. The committee examined problems and solutions regarding the introduction of the industry, and mapped out the direction of subsequent supportive measures. The

Basic Policy Subcommittee in METI, published an interim report in December 1996, where it was made clear that they were going to "work on policies to advance the establishment of ESCOs." ESCOs then came to be considered as one of Japan's energy saving policy programs.

In the promotion of energy-saving businesses like energy service companies, the energy-saving policy included in the global warming counter measures serves as an important element. Various system changes such as the Energy Conservation Law formed a common infrastructure in Japanese society to achieve the Kyoto target. In addition, the government has taken initiatives in the field of subsidization and various guideline developments.

Feasibility Study (1997)

A study group was organized for the general examination of ESCOs in 1997. Its membership totaled 233 from 208 companies and organizations. The group estimated the potential size of the ESCO market to be 4.04 million kL/year in crude oil equivalent with a corresponding value of USD 22.5 billion. It examined the feasibility of the ESCO business in commercial and industrial facilities, formalized standard contract forms and investigated measurement and verification protocols.

Demonstration Projects and Standard Contracts (1998)

Four ESCO projects were subsidized by New Energy and Industrial Technology Development Organization (NEDO) as "The building retrofit model projects for efficient use of energy in FY1998." The standard contract form was drawn up, together with the evaluation of the energy saving nature of the projects, their economical efficiency, and the measurement and verification protocol.

The standard contract form was drawn up, by examining existing work from the U.S. and other countries. The contract form prior to examination was similar to the common construction work contract, with energy saving guarantee provisions added.

The standard contract form was thought to be inadequate for dealing with unpredictable changing factors in Japan and was improved by probing the typical risks born by ESCOs and owners, and stipulating solutions to each, so as to improve dispute avoidance. Japanese entrepreneurs were basically unfamiliar with such a detailed contract. On the other hand, in public facilities, it is required at an examination stage, to provide the conditions so as to respond to various anticipated fluctuations. The standard contract reflected the contract used at the time of procurement, using ESCO projects in public facilities.

Warming Countermeasures (1998)

The Law to promote Global Warming Countermeasures (1998) established responsibilities for national government, local authorities, business institutions and people, to employ appropriate measures against global warming. It prescribed that the national government should bear the responsibility to promote measures concerning its office work and federal projects, and to draw up the plan to achieve a Kyoto Protocol target. For the ESCO industry, it became the incentive which promoted the industry's introduction in public facilities, which is considered to be the market with the biggest potential.

Measurement & Verification Protocol Guideline (1999-2002)

Research on measurement & verification has been advanced mainly by the US, which took the lead in developing the International Performance Measurement & Verification Protocol (IPMVP). Although IPMVP provided the fundamental framework and the technical contents of measurement and verification, it was revalidated and adjusted to the conditions peculiar to Japan. For the purpose of this examination, a measurement survey of energy consumption was conducted, centering on the commercial facilities over three years, from 1999 to 2002. This effort eventually developed the guidelines for the Japanese Measurement & Verification Protocol.

Establishment of the Japan Association of Energy Service Companies

The Japan Association of Energy Service Companies (JAESCO) was established as the private entity for ESCO promotion in 1999. At its initiation, only 16 organizations and individuals participated. The membership rose to 137 by March, 2008. Along with strengthening the spread of educational activities internally, it has been placing importance on the activities in other Asian nations since 2005. Activities of JAESCO are mainly focused on new market exploration and information dissemination to its members. In particular, JAESCO has been:

- *Investigating*
  for ESCO market activation

- *Strategy planning*
  for promoting the ESCO industry along with policy proposals.

- *Seminars for members*

    The explanatory meetings and opinion exchange meetings are held two to three times a year on policy, industry, finance, insurance, technology, and system issues concerning ESCOs.

- *Idea exchange sessions*

    Member study sessions are held for information exchange and improvement of skills, expanding business opportunities for the ESCO industry.

- *Exhibiting at ENEX*

    Every year, JAESCO runs a display booth at the Energy and Environment Exhibition (ENEX) held in Tokyo and Osaka by the Ministry of Economy, Trade and Industry.

- *JAESCO conferences*

    Conferences of more than 300 are held every year to discuss various problems in the industry. Lectures are given by persons involved in ESCO work with the keynote speech by a prominent, and well-informed, person.

- *Market development in the public sector*

    JAESCO takes part in the policy planning procedures to introduce ESCO projects in national government facilities by the Ministry of Environment, the Ministry of Economy, Trade and Industry, and the Ministry of Land, Infrastructure and Transport.

- *Market research*

    Market research for JAESCO members is carried out every year in which market size is estimated and project contents are analyzed. The data have been shown to be useful for policymakers as background information for policy planning and evaluation. Above all, this research has proved to be important and beneficial for the ESCO industry.

- *Asia ESCO conferences*

    JAESCO held the first Asia ESCO Conference in Thailand in 1995 and the second, in Beijing in 2007, to promote energy savings and activation of the ESCO industry in Asia. The number of participants reached 200-300.

- *Asia ESCO Association's network (AFA-Net)*
  An initial meeting was convened in 2007 to strengthen the cooperation of the Asian association people involved in ESCOs.

- *A mutual agreement with the China ESCO association*
  The statement of mutual agreement was signed between Chinese ESCO Association (EMCA) and JAESCO at "Japan-China Forum on Energy Saving and Environment" held in Beijing in September, 2007, for the purpose of supporting expansion of the ESCO industry in China.

- *Newsletters*
  Newsletters are published two or three times a year, to provide substantial information to members

Green Procurement Law (2000)

The law promoting the procurement of environmental goods by the national government became effective in 2000. Energy-saving diagnoses were specified as tasks, which should be procured by national facilities. ESCOs expect the public facilities to be a future big market, and the Green Procurement Law is expected to pave the way for ESCO introduction within national facilities.

Energy Conservation Center Support

Energy Conservation Center Japan began ESCO explanatory meetings in 2000. These meetings are held nationally in major cities every year. Pamphlets, which are designed to train and educate, and provide edited case studies, are distributed at the meetings.

Introduction of full-scale ESCO Projects to Local Authorities (2001)

The introduction of ESCO projects by Mi-e prefecture, Mitaka city and several other local authorities were the first among local authorities. Procurement rules created various restrictions to local authorities, so a full-scale ESCO introduction did not take place immediately. In Osaka prefecture, new procedures were developed for procurement allowing full-scale ESCO projects, utilizing private-sector capital.

It is common practice in Japan for local authorities to seek design and construction as separately ordered efforts. Construction, which is defined by the design, can be solely procured by competitive bid and is deemed the most economical approach. In ESCO projects, it is desirable to bundle design and construction. A new and creative procurement method was adopted by Osaka prefecture by interpreting that an ESCO project is service supply as a whole, including an energy-saving guarantee, and measurement and verification. Through the Osaka model, it became evident that ESCO project orders could be made within the limits of the present procurement rules and the approach was adopted in many local authorities. However, several problems remain in this system including risk sharing between local authorities and ESCOs, profits distribution, and other procedural problems.

ESCO Promotion as a Policy
As the result of an Energy Council pronouncement, "It is effective to develop the environment for positive and practical use of ESCO to advance energy saving as a business," it is expected that by 2010, 1 million kL/year in crude oil equivalent will be saved.

Counter Measures for Global Warming
The 2002, General Principle to Promote Counter Measures for Global Warming was made up of basic measures, taking into consideration the effectuation of the Kyoto Protocol. It was decided to perform assessment and re-examination of the General Principles in 2004 to 2007, and warming countermeasures were to be strengthened gradually henceforth. In general, the target is set for ESCOs as well as building and energy management systems (BEMS) to carry out 700,000 t-CO2 (1,600,000 kL/year in crude oil equivalent) curtailment by 2010.

With the ratification of the Kyoto Protocol by Russia in 2004, two conditions prescribed in article 25 were satisfied: 1) acceptance by at least 55 nations and 2), acceptance by developed nations whose share of CO_2 emissions totals at least 55% of that of developed nations as a whole. Kyoto Protocol went into effect as an international law on February 16, 2005, 90 days after ratification by Russia.

Preparation of Guideline to Introduce ESCO
Projects to Local Authorities (2004)
The guideline was developed to interpret procedures and promote the introduction of ESCO projects. The guideline describes introductory procedure, assessment of bidders, and standard contracts.

ESCO Project Introduction to National Facilities (2004)

The ESCO concept was introduced for the first time in National Institute for Environmental Studies in 2004, and was also introduced eventually in the Government Building of the Ministry of Economy, Trade and Industry in 2005.

The "Environmental Consideration Contracting Law" came into effect in 2008. This law revises a part of the government procurement system aiming at CO_2 emission curtailment. By this revision, the government can carry out larger scale projects, as it permits extending the debt period up to 10 years in the case of procurement via ESCOs. The guidelines and manual prepared to apply this law, specifically show a procedure to introduce ESCO projects.

ESCO Market Size

Since 2001, JAESCO has been conducting an annual market survey for its members, under the planning and guidance of the authors. Figure 6-1 shows the sales of the ESCO market. In the figure, projects accompanied by "energy saving guarantee contract" are exclusively classified as "ESCO projects" and the rest are designated "energy saving retrofits." The scale of ESCO projects is mainly analyzed from the data in 2006. ESCO projects are broken down into two categories, projects with contracts which guarantee energy savings, and energy service providers (hereafter ESP), mainly handling distributed power supply).

The scale of the ESCO industry expanded and almost doubled annually from a level of USD 7.6 million in 1998, to USD 62 million in 2001. The commercial sector, which accounted for 62 percent of the whole, grew first. The industrial sector showed gradual growth in and after 2000. During the initial phase, most of ESCOs used Guaranteed Saving Contracts (GSC) (Figure 6-2). Shared Savings Contracts (SSC) started spreading in 2000 and accounted for 48 percent in 2001 (Figure 6-2).

There were seven ESCOs in 1998, increasing to fifteen in 2001 (Figure 6-3). At this time, ESCO promotion became more active including capacity building, feasibility studies, demonstration projects, standard contracts, measurement & verification guidelines, and meetings. With respect to policy measures, the ESCO industry was positioned as one of the pillars against global warming, and JAESCO was established. Awareness of ESCOs was improving and general enlightenment activities were just getting started.

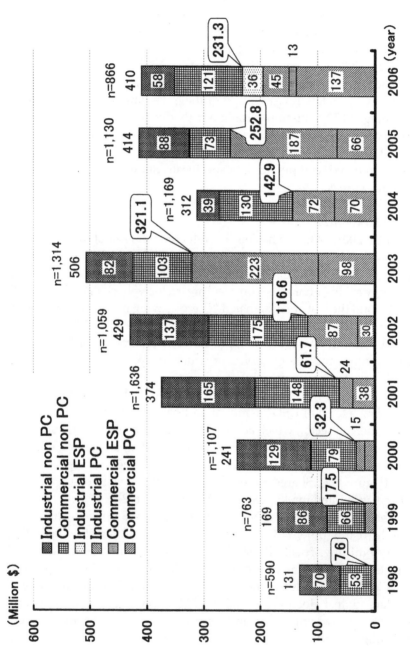

Figure 6-1. Scale of ESCO Market in Japan

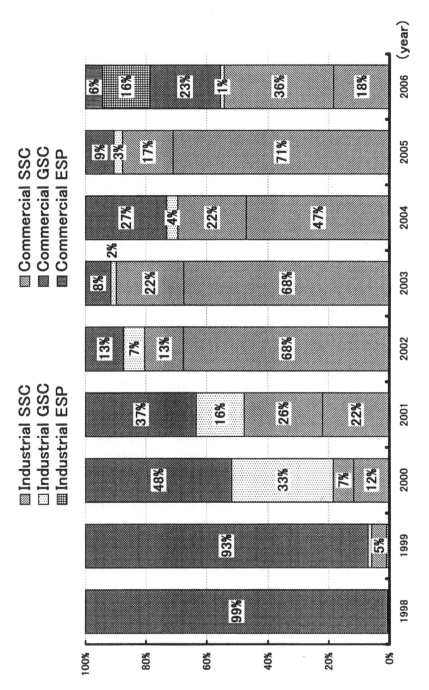

Figure 6-2. Ratio of Order Value by ESCO Contract Classification

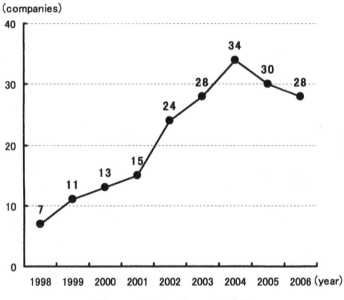

Figure 6-3. Number of ESCOs

Later Phases

The industrial market started a rapid growth in 2002. ESCO business reached the equivalent of USD 117 million, nearly double that of the previous year. In particular, the industrial sector expanded 3.7 times to USD 87 million, involving 75 percent of the industry. In 2003, this tendency became still more remarkable. The scale of ESCO market grew by 2.7 times over 2002 and reached USD 321 million. Both the commercial sector and the industrial sector showed high growth. The industrial sector marked a high of USD 233 million. Although the market grew favorably till 2003, in 2004 it fell back to USD 143 million (45 percent of the 2003 level). A depression in the industrial sector accounted for the contraction. Although not so severe as the industrial sector, the commercial sector scaled down by 29 percent over 2003.

A soaring crude oil price in 2004 encouraged more ESCO activity. The ESCO market continued to grow and reached USD 253 million.

The number of ESCO projects is shown in Figure 6-4. It has been stable at about 200 projects per year with slight fluctuations. Initially, the commercial sector was slightly larger in scale than the industrial sector. They became nearly equal in 2005.

Asia

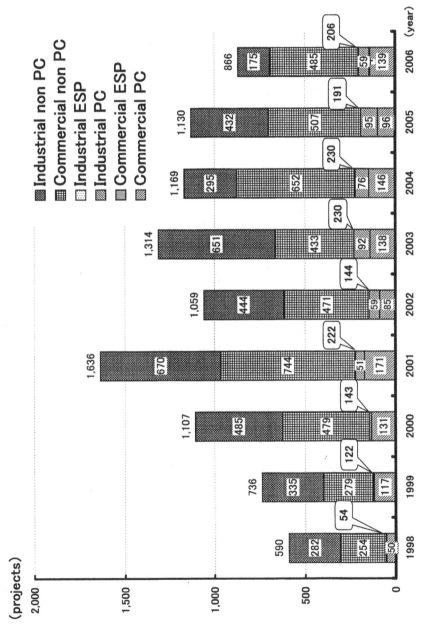

Figure 6-4. Number of ESCO Projects by Category

In the case of shared savings (SSC) in Japan, ESCOs bear many kinds of risks including the financial risk. Generally it results in short term recovery of capital. Consequently, the ratio of energy savings and the amount of investment are low. However, the amount of investment is larger in the case of shared savings than in guaranteed savings in Japan.

Many Japanese ESCOs are big companies or their subsidiaries. These are outstanding characteristics of the current ESCO industry in Japan. When further expansion of the ESCO market is pursued, it will be necessary to create financial environments favorable to owners with relatively low credit histories.

Characteristics of the ESCO project

A survey of 1,139 projects, in the period from 2001 to 2005, was conducted by JAESCO. Over this period 40 percent of the contracts (1012) contained energy-saving guarantees (called PC projects here and the remainder of this section on Japan) and the rest, ordinary energy saving retrofits (non-PC).

With regard to the commercial sector, total floor area of PC contracts was 36,500 m^2 whereas 28,800 m^2 for non-PC projects. The total floor area of PC projects in the commercial sector is classified into four groups, namely hospitals, hotels, and commercial institutions with 43,800m^2, offices with 36,600m^2, schools and research organization with 25,600m^2 and social welfare institution with 17,300m^2. Of great interest is the finding that projects of the large scale are numerous (Figure 6-5).

Ratio of Energy Savings

When classified by contract category, SSC marks higher energy savings than GSC in both the commercial and the industrial sectors. With respect to shared savings, since ESCOs offer funding necessary to implement the project, short term recovery of the fund is requisite for risk aversion purposes. Financial institutions also attach high value to the credibility of ESCOs, the loan to small or medium-sized ESCOs tends to be with short repayment periods. In Japan, however, SSCs are implemented with a high ratio of energy savings and are of large scale. This stems from the characteristic of Japanese ESCOs, that most of their customers are high-rated companies and the ESCOs themselves are also big companies or affiliates of large companies.

On PC projects in commercial institutions, the energy saving rate is the highest in hospitals with 16.3 percent, followed by social welfare

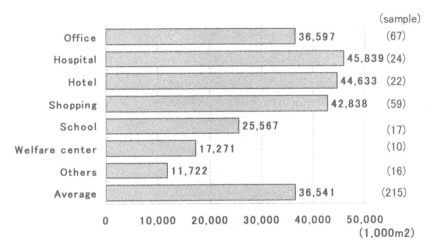

Figure 6-5. Total Floor Area of PC Projects Investigated

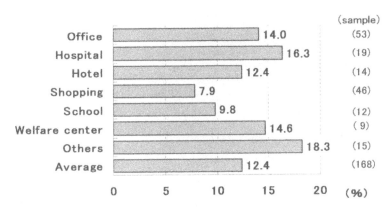

Figure 6-6. Ratio of Energy Saving by Facility in the Commercial Sector

institutions with 14.6 percent, offices with 14 percent, and shopping institutions with 7.9 percent. As in the case of hospitals, a high ratio of energy savings is attained where large heat demand and long working hours exist; hence, large scaled investment is easy to be adopted with no payback difficulties.

Contracts

The average contract amount of PC projects is larger in industrial facilities (USD 2.55 million) than in commercial facilities (USD 1.1 million) since the projects in the industrial facilities contain many retrofit projects of heat source machines including cogeneration systems. The higher ratio of energy savings (Figure 6-7) and the larger scale of PC projects creates a predominance of PC type ESCO projects.

If classified by contract category, shared saving payout years are about two times longer than guaranteed savings projects in both the industrial and the commercial sectors. Shortness of the simple payback period for shops is attributable to low ratio of energy savings and small scale of investment.

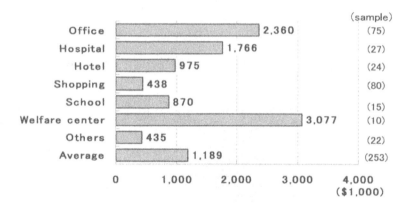

Figure 6-7. Contract Amount by Sector and by Contract Category

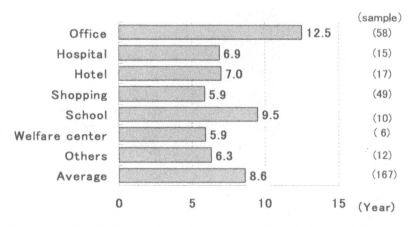

Figure 6-8. Simple Payback Period PC Projects Classified by Business Type

Technologies Adopted

Adopted technologies are shown in Figure 6-9 (the commercial sector) and 6-12 (the industrial sector). Air-conditioning pump fan's inverter introduction is the most popular technology adopted in the commercial facilities (53 percent of the total). Inverter lighting comes next (30 percent). Lighting construction is also performed in almost all facilities, including HF inverter (19 percent), electric bulb type fluorescent lights (12 percent) and high luminosity guidance lights (12 percent). On air-conditioning, subsequent to a pump fan's inverter introduction, CO_2 control is as high as 22 percent. On heat source equipment, renewal of freezer (14 percent) and introduction of cogeneration (10 percent) are typical building energy management systems and are adopted by 15 percent.

Cogeneration introduction is the technology most widely utilized in the industrial facilities. It is adopted in more than one third of all projects. With regard to heat source equipment, renewal of a chiller (16 percent) and

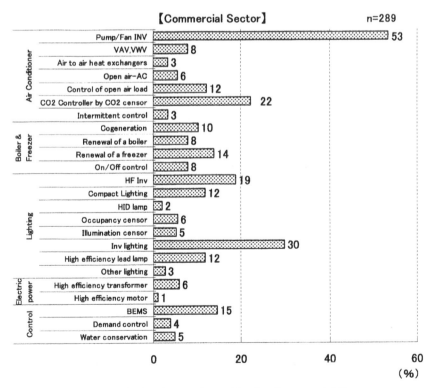

Figure 6-9. Technology Adopted in the Commercial Sector

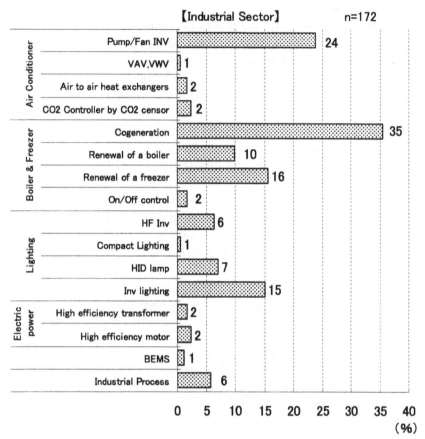

Figure 6-10. Technology Adopted in the Industrial Sector

renewal of a boiler (10 percent) are adopted somewhat more frequently in the commercial facilities.

Conclusions

Since the introduction of the ESCO industry in 1996, the industry has been regarded as a leader among the measures against global warming in Japan, and it has continued to receive governmental support.

In Japan, the private sector comprises most of the ESCO market. The market grew to the equivalent of USD 231 million by 2006. The rapid early growth was mainly attributed to the expansion of the projects with shared savings contracts in the industrial sector. The introduction of the oil cogeneration system, with large-sized investments, has contributed to the growth. In 2004, introduction of oil-cogeneration systems decreased due to the oil price hike, hence, the industrial sector market contracted.

Since the number of ESCO projects have stayed around two hundred a year, change in the market scale has been attributable to the fluctuation of the construction scale of each project. Although the industrial sector recovered in 2005, marking record high of USD 253 million, it once again shrank in 2006 to USD 231 million. Despite market fluctuation, the commercial sector expanded steadily to a record high of USD 150 million in 2006, and is expected to lead the future ESCO market.

Shared savings contracts account for as much as 76 percent of the market. The macro-scale of the projects is particularly outstanding in shared savings contracts. ESCO projects including energy saving guarantees averaged USD 1.12 million per project, 4.2 times as large as the common types of energy saving retrofits which average USD 0.27 million per project. It is considered to be a proof that ESCO projects have been accepted as a tool to implement large scale energy saving retrofitting work. Project size of shared savings contracts averaged USD 2.73 million, which substantially exceed that of guaranteed saving contracts at USD 0.40 million. The rate of energy savings in ESCO projects is about 12 percent. Simple payback periods are about nine years as a whole, while the simple payback period for shared savings contracts is about 11 years.

Most ESCOs in Japan are subsidiaries of large companies. The ESCO industry of Japan is still in a developmental stage, and there are future possibilities for project deployment in an extensive field, including small or medium-sized customers. Although the rise of the oil price in 2004 and afterwards caused a reduction of the market temporarily, it will become a factor of market expansion in the long run. Moreover, as the necessity for measures against global warming becomes strong and regulations are toughened, the ESCO industry is expected to expand further.

THAILAND

ESCO pioneers in Thailand, Mr. Arthit Vechakij and Mr. Ruamlarp Anantasanta provide the following insightful view of the development of the ESCO industry in Thailand. Of particular interest are the familiar concerns of end users as to why they do not implement energy conservation and efficiency measures.

For ESCO business developers and ESCO business operators, Thailand offers a really good case study, especially for developing countries, South East Asian countries, or even most Asian countries. The Thai authors have tried to minimize the introduction of names/terminologies that are specific to Thailand, but some are required by the context. If you still feel it is too much, please kindly accept our apologies.

A summary of the progression of ESCO development in Thailand is presented in Table 6-1.

Early Motivation of the Energy Efficiency Market

In 1992, the Thai government announced the Energy Conservation Promotion (ENCON) Act to be used as a regulatory framework for energy conservation action. With this act, one of the early, and well-known laws announced, was "Target & Plan." Directed at the high energy consumption users (described as designated factories or buildings), this law required them to regularly measure their energy use, perform a detailed audit, and most importantly, set their targets of new, and decreased, energy usage, as well as develop a plan on when, and how much, the level of energy would

Table 6-1. Progression of ESCO Development in Thailand

1999: The birth of ESCO business in Thailand … the Commencement Day
The Thailand ESCO Pilot Project 1999; sponsored by the World Bank and conducted by the Ministry of Energy and EGAT (Electricity Generating Authority of Thailand) with the core intention to establish an ESCO business ❖ Thailand Energy Efficiency ("EE") market at that time ❖ The companies who were interested in Thailand ESCO business at that time ❖ Finally, the ESCO Project resulting from the ESCO Pilot Project
After that; the first few years of ESCO business in Thailand … the Stumbling Blocks
❖ The perception of ESCO market itself and as part of the whole energy efficiency (EE), renewable energy (RE) and alternative energy (AE) market in Thailand ❖ Thai government support policy and programme on EE/RE/AE/ESCO: What is <u>HOT</u> and what is <u>NOT</u>!
Since 2005: The urge for Energy Savings in Thailand … the New Beginning
❖ The increase in energy prices; especially fuel oil and electricity (and even transportation fuels) ❖ The HOT support policy and programme at the hot time by Thai government! ❖ Perception of financial institutes in Thailand on EE/RE/AE/ESCO and the changes ❖ ESCO shared savings in Thailand
Yesterday-Today-Tomorrow: The ESCO Business in Thailand … the Outlook of Challenge
❖ The Barriers, and the Opportunities! ❖ The future of ESCO in Thailand
1999: The birth of ESCO business in Thailand … the Commencement Day

gradually be cut. This plan also addressed the specific activities high energy consumers would use in order to achieve such reduction. On the industry side, any factory which consumed more than 1 MW of electricity, was considered a "designated factory." On the commercial building side, any building consuming more than a specified amount of electricity was considered a "designated building."

The target and plan approach sounded like a great model; however, reality seldom reflects ideology. Figures recorded by credible sources show that though part of the projects, as advised by the Target & Plan report, were implemented and could, therefore, be defined as a 'success,' there were still many good energy conservation projects left not yet implemented! The Target & Plan required that a Target & Plan *report* be submitted. The follow-up as to how the plan was to be executed was not enforced.

Barriers

The research has addressed the most frequent reasons why energy users in Thailand decide not to implement energy conservation or energy efficiency measures. Potential customers are surrounded by confusion as to:

1. Which project should they do first?
2. Which technology option should they proceed with?
3. Which size project is really suitable to them?
4. What if the energy savings project does not save?
5. What if the energy savings project does not save as much, or do it as soon, as the feasibility study suggests; i.e., how credible is the feasibility study?
6. What is the likelihood that energy users will invest in project development; e.g., hiring someone to conduct feasibility study?
7. What if the energy savings project gets stuck in the middle of implementation and never has a chance to pay off?
8. Is there any clear energy savings expert out there to assist the project?
9. The project return on investment is not considered attractive enough to proceed when compared to their conventional business capacity expansion or capability (product line) extension. This impacts budgeting for both capital and resource requirement.
10. To most industry, but not all, energy is typically not ranked as a high-attention area, compared to other raw materials, as energy prices are still considered low.

11. Investment environment and credit—beliefs held by a company or its financial institution were that granting a loan would not be a good investment during the belt-tightening economics after the "1997 Tom Yum Goong." *(The collapse of Thai baht in 1997, which triggered Asian economics to slump in the same year.)*
12. Greater than all the concerns stated above, is the reluctance of a company to invest in the energy savings project when the first project did not prove itself to be a good investment. After granting the first project and waiting until it was implemented, the management still did not have a chance to see any actual savings, as the in-house engineers were not equipped with the needed M&V skills to document the savings.
13. In some projects, the actual savings advantage was lost in the lackluster performance of O&M personnel.

From both inside-out and outside-in vision, the Thai government—under the Department of Alternative Energy Development and Efficiency ("DEDE"), which is directly accountable for Energy Conservation under Ministry of Thailand ("MoEN"), and the World Bank GEF (Global Environment Facility)—began to see the needs to resolve energy users' concerns by introducing the ESCO business model into Thailand.

Working from both sides, DEDE, with the World Bank GEF's assistance, has been developing an ESCO industry in Thailand. EGAT, the Thai utility, also joined as the Implementing Agency (IA) of the "Thailand ESCO Pilot Project." The term "pilot" was used to reflect the trial of this new business model, as the procedures introduced a never-before-seen legal framework and financial risk management through the performance contractor. All projects from the Thailand ESCO Pilot Project were meant to demonstrate how this ESCO-type of project would work, to serve as a showcase to facilities, financial institutions, and all other relevant agencies.

Companies with an Early Interest in the ESCO Business

The 1992 announcement of the ENCON Act was considered the very beginning of an energy conservation industry in Thailand. It has not yet yielded huge savings as specified in the Target & Plan requirement. Once the news circulated about three reputable organizations trying to roll out the ESCO industry here, there was wide interest among a variety of existing business players and entrepreneurs. They generally fell into the following categories:

- Local (Thai) company in a energy related industry; e.g. consultant (energy audit, energy efficiency), equipment manufacturer, turnkey contractor;
- Entrepreneurs interested in the energy conservation industry and professionals in energy or related fields;
- Consultant companies in energy industries from abroad;
- ESCO companies from abroad; and
- Equipment manufacturer related to energy/energy savings industry from abroad.

For most of them, it was not only to take advantage of the opportunities in the emerging energy savings business in Thailand, but to also use the driving force from DEDE, EGAT, and World Bank GEF as they provided an "environment" where such activities could take place, including:

(i) The use of credibility from DEDE and EGAT to attract customers into an ESCO pilot project;
(ii) The use of sponsorship from DEDE and EGAT, in terms of effort and campaign PR, to attract customers into an ESCO Pilot Project;
(iii) The use of credibility, sponsorship, and support from DEDE, EGAT, and World Bank GEF to bring together demand and supply; i.e., customers and a business entity employing the ESCO business model in a risk guaranteed energy efficiency project; and
(iv) The use of supporting grants from World Bank GEF to fund the ESCO activities in project development; i.e., preliminary study—project Identification, investment grade audit (IGA).

The support from World Bank GEF through DEDE/EGAT for the ESCO Pilot Project went beyond the grants for preliminary audit and investment grade audit (IGA), to the very intensive training programs provided by experts on all areas related to ESCO business. These programs covered energy performance contracting (EPC), measurement and verification plans (M&V), energy conservation measures (ECMs), preliminary and detail audit procedures, etc. Moreover, through their reputation, EGAT's action as the promoter of this programme was valuable in helping to develop initial trust between new customers and new ESCOs.

This campaign was begun with more than ten companies, but later became four companies trying to work their way into becoming ESCOs.

Results from the ESCO Pilot Project

With the involvement of EE and ESCO experts in this green field operation, and with important sponsorship, it would seem that the ESCO pilot project would run smoothly and be a success with a number of projects. However, while we all thought that the ESCO-guaranteed savings could lubricate the barrier of energy efficiency project, *deal closures* between customers and ESCOs also carried another significant barrier of their own!

Unlike America and Europe, in Thailand in 1999 (similar to other developing countries), the business firms were not familiar with non-conventional outsourcing, especially for intangibles such as knowledge, know-how, consultancy, and risk. It was okay to pay someone/some company to perform some work they could not really do internally. However, if they already had the existing staff nearby; e.g., engineers, they would not want to pay the added expense for the energy efficiency expertise, which they believed unnecessary. That was the first area of reluctance towards using ESCO services. It was difficult to accept that in-house engineers were not equal to energy efficiency experts. Further, the ESCO field doesn't cover just the engineering point of view, but that is how the potential Thai customers perceived it.

The second barrier came under "the difference in corporate environment." ESCOs here share this concern with other consultancy services like business and IT. Table 6-2 documents these cultural differences.

The details in Table 6-2 explain why the consultancy business in Thailand has difficulty as their value cannot be easily accepted since it offers a service in an area not yet assigned to anybody. Even with a large

Table 6-2. Cultural Differences

Compared Area	US/Europe	Thailand (at year 1999)

Firm Structure	Well-structured corporate	1. Non-listed Company
		   - Family Company
		   - Founder tries to fade off
		2. Listed Company

Executive & Management	Professional Staff - Understand company area of expertise And understand reason and value for outsourcing	Owner themselves - Believed everything can be done internally - Tended to think of cash out of pocket at expense, rather than worthwhile investment
Staff	Employee - Who well understand Scope & Responsibility (and prefer if the company hires expert to help)	Employee - Who could fear that if outside experts worked their way, it would imply their inability and shake job security.
Perceive value of expert consultancy & paying for knowhow	Yes	No - Pay for equipment (tangible assets) only

corporate firm—and a vastly negotiated deal—the disadvantages of being their first footprint in the firm for a future deal is a struggle.

The third barrier was under the same cultural umbrella, but this one was "the difference in the legal environment." This was not about the governing law of Thailand or anything that was quite different from US or Europe, but it was about how contracts between two parties would be opened, discussed, understood, negotiated, bound, followed, and exercised. When it came to performance contracting, this was even harder because there are more risks associated. In the "Third World," customers still highly prefer 'loose' contracts.

Combining these three major barriers with the fact that:
- Performance contracting was very new and innovative to Thailand;
- The highly experienced international ESCO and energy industry firms had to sustain some cultural differences;
- The inability of the newborn local firm to comprehensively conduct ESCO work, or even to explain ESCO procedures and energy performance contracting; and
- Commercial lenders; i.e. banks, did not easily understand that the cash flow from energy savings, under the ESCO risk management, could be trusted enough to repay the loan.

All relevant parties (World Bank GEF, DEDE, EGAT, ESCOs, and customers) had to cruise through such barriers. There was a back and forth of trying to trim down the interested pairs of customers and ESCOs from more than ten, down to the final four. After a long journey, finally, one firm survived. The identified energy efficiency project was financed through the commercial lender, was implemented, and has been in operation since October 2003 under the target savings guaranteed scheme. That firm is the author's Excellent Energy International Company, Ltd.

The pilot project customer is Bangkok Produce Merchandising PLC ("BKP"), a subsidiary of Charoen Pokphand ("CP"), one of the top firms in the Thailand food industry. The energy conservation measures selected were the gas turbine cogeneration system fueled by natural gas, which can simultaneously produce 4.5 MW of electricity and 12 tons per hour of steam. The total investment was approximately USD 4.5M with net savings of USD 1 M per annum at that time.

It is worthwhile to analyze why this ESCO Pilot project succeeded. It did so for the following reasons.
- The selected project was quite big. The potential savings provided

sufficient room to draw attention from customer's staff, engineers, management, executives, and even their board of directors and founders. Getting approval for this kind of project in Thailand is really time consuming and provided no room for error, misleading, misunderstanding, and distraction!

- The customer is a very big and creditworthy company. Therefore, the project did not get much resistance from the financial institutions to provide the loan. In fact, they were even bidding for this project. The fact that this project had its risk minimized through the ESCO model may not have played much of a role.

Still, the project was almost not carried out as its benefits were still lower—compared to their standard business returns. The fact that most companies in Thailand cling more to the IRR could easily have made this low-risk project overlooked. Finally, DEDE & EGAT had decided to absorb the interest expense of the project; thus, the customer ended up needing to repay the loan principal only.

Through these success factors and significant supports, this ESCO project could not deny its birth, so finally a true ESCO Project resulted from the Thailand ESCO Pilot Project.

Perception of the ESCO Market

By 2001, the market had reached the stage where an EPC was accepted by stakeholders (customers, ESCOs, banks) and there was a loan approved by the bank for the ESCO EEI project. The ESCO business was ready to roll based on this commercial activity with the supporting precedent case. ESCO companies in Thailand had started approaching new prospects, but history kept repeating itself. All the barriers that ESCO companies had to face during the Thailand ESCO Pilot Project were still there! They had not yet been removed. Though some guards were a bit lower due to the demonstration project, it was still too high to jump across. Especially, where there was no such catalyst as a pilot program.

During this period, however, a number of EE projects were implemented, but at a slow rate. Mainly because the leading industries had already rebounded from the economic crisis and some of them, the visionaries, had started contemplating ways to strengthen their business through energy conservation actions. Yet, the intrinsic barrier for Thai customers to adopt an ESCO scheme never vanished. While ESCOs offer risk shedding: mechanisms for them, the clients seemed to believe that they could comfortably handle the EE technical risk by themselves; thus, ESCOs did not add much perceived value. Or, if they foresaw some risk, they still took the mind-set of "high-risk, high-return," rather than switching to the ESCO value proposition of "low-risk, a little less return." In short, major customers were not willing to pay guaranteed/shared savings fees to ESCOs.

Mostly during this period, Thai customers decided to work on their own with in-house engineers, or purchase relevant EE equipment. They were content to get what they believed was "free consultancy" provided by equipment suppliers. If the EE project was big, customers were content to settle for a direct handshake between themselves and the turnkey contractor.

At this point in time, renewable energy (RE) projects were still mostly talk, and any projects were in the far future. The words *ethanol, biodiesel, solar cell,* and *wind turbine* were not widely spread.

Alternative energy (AE) projects were more common. The market penetration of natural gas pipelines by PPT (the sole state enterprise for refinery and natural gas) used two selling points: first, AE was a little cheaper than heavy fuel oil and LPG. Second, the wording of "natural" gas provided a sense of clean energy. The waste from agricultural business, *biomass*, started coming into the alternative energy picture.

The customers, who chose the ESCO performance contracting scheme during this period, were the younger generation of executives with study/work experience in developed countries. They tended to understand the value of outsourcing, knowledge, and risk shedding in their non-core business.

The other type of project that had appeal at this time was innovative technologies from abroad. The ESCO was looked to for the expertise needed.

Not many EE/ESCO projects were widely implemented during this *awfully tough* period, but they represented the technology and were ESCO showcase projects. As EE/ESCO projects began booming, there was a growing support for ESCOs to present their energy conservation measures as "proven technology that exists in Thailand."

Thai Government Support Policy and Program

What is HOT and what is NOT! The key strategic direction of the Thai government on energy efficiency improvements consisted of: (i) improving EE in transportation; (ii) improving EE in industry; and (iii) an awareness raising campaign. A conducive environment for EE investments was created and is displayed in the Table 6-3.

What worked well for ESCOs during this period was the replaying of the past measures, which had worked well during the ESCO pilot project. For instance, the technical assistance, which is the program that the governments sponsored, allowed ESCOs to perform free preliminary audits and consulting for the factories. This lubricated ESCO marketing activity when both parties (potential customers and ESCOs) did not have to absorb the fee. This made ESCOs able to present the project in a more attractive manner by suggesting the ECMs right away, before entering into any complex contracting framework.

The low interest loan also accelerated energy efficiency in Thailand and has been successful all along. On the financing side, this approach gives more benefit to energy efficiency activities compared to other "business-as-usual" activities.

These support activities offered a strong foundation, which helped prepare for the time EE and ESCOs would be recognized for their importance by industrialists in Thailand.

The Urge Toward Energy Savings… The New Beginning

As mentioned earlier. Several conditions have helped ESCO growth:

Table 6-3. Energy Efficiency Support

Program Area	Measures	Notes
Incentives & Subsities	Low interest loan	HOT
	Grant for small investments	
	Tax incentives	EE, RE & ESCO
Technical assistance	Energy Clinic	
	Workshop & training	
	Free energy audit and consultation	Useful to ESCO
Regulations	Energy codes and standards	
Raise up confidence	Promotion campaign Demonstration project Promotion of ESCOs	Awareness
Perfect information	Directory of technology providers List of experts	

- The high rise of energy prices; especially fuel oil and electricity (and even vehicle fuels);
- The HOT support policy and programme at the hot time by Thai government; and
- Perception of financial institutions in Thailand on EE/RE/AE/ESCO and the changes.

Interestingly, these three conditions created a never-before-seen positive domino effect, and created the basis for the surge for a new beginning of an ESCO industry in Thailand.

Since 2005, there have been rising oil prices all over the world. In Thailand, it impacted every sector—industrial use, home use, and vehicle use. This surge is expected to persist longer than any previous oil crisis, as it has been triggered by the factors of emerging of China industry and the real depletion of this natural resource, not market intervention as in the past; e.g. by OPEC. This has really opened "customer's eyes." Top executives of leading firms in Thailand are deciding to seriously proceed with energy conservation projects, and are declaring,

> *"We need a strong foundation. Every cost has to be optimized. This specifically includes our energy costs, which have been growing most compared to other costs and even revenue! Our efficiently managed energy really gives us a competitive advantage, globally. Looking around, it is not easy to increase the market size and revenue. Maybe it is best to increase our profit by seriously cutting cost."*

The Ministry of Energy has continued its promotion of ESCOs and overall energy conservation, efficiency, renewable, and alternative energy support. The first Thailand ESCO Fair was held in Bangkok early FY 2005. At that time, it was more of a "business-to-business" and association event. It was the first time MoEN, the relevant government organizations, and ESCOs grouped together to showcase to the public the existence of an industry to help customers and the country achieve energy savings. Success cases were presented. The fair received good coverage in the press, but interest from prospective customers was still not much. Anyway, it was the fireworks and showcase on ESCOs for the public, media, local banks, and other organizations.

The "Low Interest Rate Loan for Energy Conservation" programmes proved to be very successful through the fast granting of 2,000 million

Baht (USD 50M). MoEN further supported the second round. The third round for this low interest rate loan is now underway and is providing direction to local banks to blend their portion of direct loans into the mix. As in other countries, the subsidy has been reduced little by little as the market—both customers and banks—has become more active on its own.

The Low Interest Rate Loan for Energy Conservation programmes has pulled local banks into loving EE/RE/AE/ESCO projects. During this global economic recession era (everywhere but China!) and energy crisis, local banks see great opportunity to flagship energy savings project for customers (existing and new) to use their loan service. Along with the continuous promotion by Thai government, local banks have gained a greater understanding of the associated risks in energy savings project. Further, they understand that they are low and controllable risks.

Due to the fact that Thailand is a developing country, thus not decentralized yet, most local banks have their headquarters in Bangkok. Fortunately, this means it is always the same energy financing team, which reviews the contracts, and have a point of contact with MoEN, to work through the bank's benefits from the low interest loan program. Having to understand the client's EC projects, the team can appraise the case to both MoEN and the loan committees of their banks. This indirectly gives great experience to our local banks in becoming familiar with our EC projects. In short, their perception has been changed gradually to positive.

The 2nd Thailand ESCO Fair took place in 2007 in Bangkok. This time, most local banks also joined the fair. This was seen as evidence that they were ready to support customers, who wished to work with ESCOs. The fair also drew high interest from customers. The key additive was that MoEN assigned Thailand FTI (Federation of Thailand Industry) to be the fair organizer. FTI utilized its network to effectively have provincial FTI units communicate to members. The preparations also had local banks and ESCOs participate in the planning. 3,000 people came to the fair, including engineers, manufacturers, senior management and financiers.

In addition to these two ESCO specific local events, there have been plenty of international and local events about energy efficiency and renewable energy held in Thailand. These events usually cover the context of EE/RE/AE/EC and ESCOs. Customers (at all levels) increasingly come to these events every day. Many leading firms in Thailand have finally come to realize that they need to become more socially responsible, which creates a never-before-seen favorable environment for EC/EE/RE and ESCOs in Thailand.

Shared Savings

Since the birth of ESCO industry in Thailand, shared savings has had two specific barriers above all the other barriers. The first barrier is the clients' decision to invest themselves once their confidence to proceed with an energy conservation project is established. It has actually been very uncomfortable for clients to choose to have their projects financed by someone at a higher cost of capital than loan interest rates. The dilemma here is that the client would like to engage in ESCO shared savings service, mostly because they could not get the loan directly from the bank; and mostly due to their high credit risk. And ESCOs could be very reluctant to enter a contract with a high credit risk customer. While the ESCO may aim to propose their shared savings service to the very good credit customer, once successfully convinced about the project, the customer may break with the ESCO and take the loan/cash to invest themselves.

Second, it is not easy for expertise-based ESCO companies to get the loan from a local bank to conduct shared savings service due to the bank's perceived credit risk. After the first few projects, an ESCO reaches its credit limit and cannot ask for more loans without increasing its company's equity size.

However, the ESCO companies who are international equipment manufacturers have fewer problems as their company size easily allows this financing scheme. In fact, it comes very close to equipment leasing model.

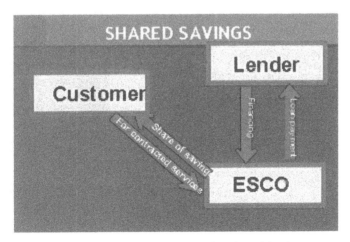

Figure 6-11. Contracted Services in Shared Savings
Source: Hansen Associates

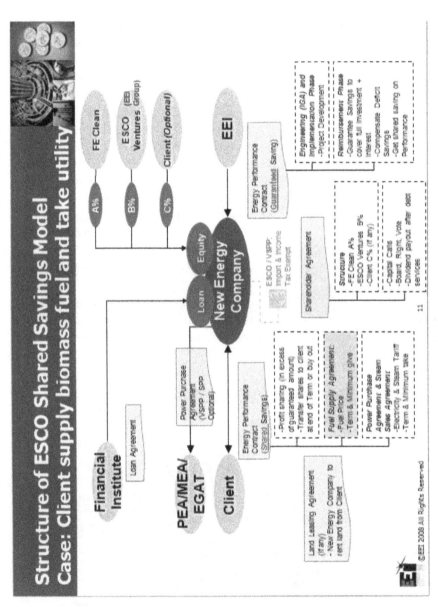

Figure 6-12. Example of ESCO Shared Savings Scheme

2005 was the wake-up time for all, and it included the shared savings scheme. Since 2005, based on economic environment, Thai companies have chosen to focus more on their core business, on both investment and resources, and are reluctant to self-invest (or use up their credit line) in non-core business. Here shared savings has good potential. The ESCO shared savings model through the SPC (special purpose company) is becoming popular in Thailand. This is especially true when the project is large size like a cogeneration plant. Using this approach, the ESCO shared savings model is often perceived by the customer as the B.O.O.T. (build-own-operate-transfer) approach. So, it is convenient to set up an SPC to handle all transactions related to the project. Moreover, the SPC somehow eases the local bank to view this project more as *Project Financing*, not *Corporate Financing*. In theory, it is not, but recipients are happy with it. Most importantly, an SPC is eligible to obtain tax privilege from the Thailand Board of Investment (BOI), which will deliver an even better project yield.

Project investors could be the ESCO company itself, or a private equity fund (PEF) in the area of energy conservation, EE, AE, RE. Sometimes, clients co-invest in the SPC. The PEFs active in Thailand are mostly come from abroad. Local efforts to form PEFs are emerging, but are not yet as fully developed as those with managers from US, Europe, and Asia.

ESCO shared savings demonstrates an extreme environmental readiness for the ESCO Industry. It requires great understanding from the bank to assess risks associated with each party. It requires customers willing to perform an energy conservation project with an ESCO. It also requires the ESCO to perform risk identification and management effectively so it can offer a bankable project that gets the most favorable terms. Under SPC scheme, it requires investors to believe in the customer creditworthiness and to have a good faith in the Thailand investment environment; i.e. economic, political ...

Barriers and Opportunities

The barriers have been gradually eliminated over time. However, those barriers are considered external and environmental ones, which change with time or when new factors or market transformations take place. An ESCO company must also be equipped to deal with intrinsic barriers in Thailand.

A multi-disciplinary approach is strongly required in the ESCO business, including:

Figure 6-13. ESCO Success Factors

- Engineering/Technical
- Marketing
- Follow updated EE/RE/AE Technology
- Project Management/Contract Management
- Deep understanding of EPC/Legal Knowledge
- M&V Skill
- Business Development
- Financial/Accounting
- O&M

Due to very high, and long-term, responsibility required under a single performance guaranteed contract, there is no doubt that an ESCO must have all the above skills available. One option is to develop a reliable and accountable partnership network. *Outsourcing is a mandatory strategy of every lean company.* While one cannot outsource its business core function, an ESCO, positioned to be a one-stop-service through big scope coverage, often has difficulty determining its core function.

To meet the broad range of needs, an ESCO should:
- Be able to select the proper customer and determine the proper energy conservation measures on the premises of that customer;
- Have very strong supportive founders/shareholders as this business will take quite a long time from grass roots starting till the stable situation is attained under a very high risk business operation;
- Recognize a reputation must be built and carefully maintained with very a broad business network, including:
  — financial institutions;
  — energy, industrial, and government sectors;
  — technology suppliers/contractors; and
  — customers, etc.;
- Have great internal capabilities as staff is the only "company asset" since ESCOs are service companies; thus people are the real key to success.

Because of these demanding requirements, the opportunity for the ESCO business itself will always be there.
- Few competitors. *All combined barriers prove themselves to be very tough to the 'New Entry;' so the ESCOs are fighting more against "substitutes"; which is actually an effort to leverage up our industry!*

[For number of ESCOs in Thailand and their respective specializa-

tions, please refer to www.thaiesco.org which is hosted by DEDE-MoEN and currently maintained by Thailand FTI.
- Many good customers and strong energy conservation measures are available in Thailand, especially under the current energy price crisis.
- Relentless good support programmes from Thai government on energy efficiency development and ESCO business development exist.
- The level of awareness of business owners has been continuously increasing throughout the past ten years.

The Future of ESCOs in Thailand

The future of ESCO in Thailand is apt to be similar to some other countries, as it will be based on our business culture and economic situation. The strength of the customer business is the vital key for the growth of ESCO business. Moreover, the ESCO business can be applied to many other areas where performance guaranteed contracting will fit, and add its unique value to the demand-supply gap. Nevertheless, the energy sector is expected to remain the main driver to the future of ESCO business. So far, steadily increasing energy prices directly play a major role in ESCO business growth and the situation is broadly expected to be maintained for several more years.

MALAYSIA

This succinct summary of the status of the ESCO industry in Malaysia is provided us by Dr. Jim K.Y. Lim, who is a pioneer in energy software and hardware development.

Situated in Southeast Asia, Malaysia, with an area of 329,750 sq km (127,317 sq mi), consists of two non-contiguous areas: Peninsular Malaysia (formerly West Malaysia), on the Asian mainland, and the states of Sarawak and Sabah, known together as East Malaysia, on the island of Borneo. Comparatively, the area occupied by Malaysia is slightly larger than the state of New Mexico.

Malaysia claims several atolls of the Spratly Island group in the South China Sea. The claim, in a region where oil is suspected, is disputed by China, the Philippines, Taiwan, and Vietnam. Malaysia's capital city, Kuala Lumpur, is located in the western part of Peninsular Malaysia.

Under the Eighth Malaysian Plan and the Ten Year Outline Perspective Plan (OPP3), energy efficiency (EE) has been recognized as an important measure to increase the competitiveness of the country's goods and services. Steps to promote and integrate EE into the energy scene have been incorporated in these plans.

One of the key plans under OPP3 is The Malaysian Industrial Energy Efficiency Improvement Project, which was launched in 2002.

The Malaysian Industrial Energy Efficiency Improvement Project

The Malaysian Industrial Energy Efficiency Improvement Project (MIEEIP) was developed to remove barriers to the efficient use of energy by the industrial sector. The project is also designed to facilitate the reduction in greenhouse gas emissions by that sector.

The MIEEIP moved to meet its objectives by creating sustainable institutional capacity to provide energy sources, and an effective policy planning and research framework. By promoting energy efficiency in the industrial sector, the project also aims to reduce the emissions of greenhouse gases that contribute towards global warming.

The project was jointly funded by:
- The Government of Malaysia (USD 6.3 million);
- The Global Environment Facility (USD 7.3 million);
- The United Nations Development Programme (USD 300, 000); and
- The Malaysian private sector (USD 5.26 million).

The Ministry of Energy, Water and Communication is the executing agency for the project, which is implemented by Pusat Tenaga Malaysia.

Project Objectives Are:
- To implement appropriate measures, such as demonstrating the effectiveness of energy saving technologies and financial incentives for the manufacture of energy efficient industrial equipment.

- To provide highly skilled energy audit and engineering services, project financing, training, and information to plant managers and operators.

- To implement energy efficiency and other large-scale efficiency programmes by strengthening the institutional capacity for energy programme design and implementation, monitoring and evaluation.

- To build the capacity of existing organizations to provide energy management advisory services, and energy engineering and services.

- To create sustainable follow-up programmes after the completion of the project that will build on the achievements and experiences gained, and where necessary, improve the activities and deliverables.

Industrial Sectors

The activities under the MIEEIP were implemented for the eight industrial sectors selected for the project: cement, ceramic, food, glass, iron & steel, pulp & paper, rubber and wood. These sectors were chosen using a selection process based on information derived from the Asian Development Bank Project Report of 1994 and a questionnaire prepared and circulated by the project proponents to energy intensive industries. Several other studies were also undertaken by the Ministry of Energy, Water and Communications in collaboration with the Japan External Trade Organisation (JETRO) regarding energy conservation opportunities in some of the energy intensive industrial sectors.

In 2005, three additional industrial sectors were selected for the audit programme. PTM and ESCO companies jointly carried out audits in the textile and oleo-chemical industries.

MEIIP Program… Benchmarking

The overall objective of this programme was to establish and develop energy-use benchmarks for the eight industrial sectors, which could

Table 6-4. Project Stakeholders

Core Stakeholders:	Beneficiaries:
Economic Planning Unit	Federation of Malaysian Manufacturers (FMM)
Ministry of Energy, Water and Communications	Industries
Energy Commission	Energy Service Companies (ESCOs)
Pusat Tenaga Malaysia	Research Institutions (Universities)
United Nations Development Programme	Sirim Bhd
Global Environment Facility	National Productivity Corporation (NPC)

be used by the industries as a guide in their energy efficiency and energy conservation (EE & EC) efforts.

To date, the programme has set up a data collection system for energy benchmarking and installed an energy-use database. Its ongoing activities include:
- Establishing industrial energy-use benchmarks; and
- Establishing a system for the dissemination of end-use benchmarking information to industries,

Audit Programme

Energy auditing is a proven effective energy management tool and has been practiced by energy professionals in Malaysia since the 1980s. However, many industries are not aware of the benefits of energy audits. Those who do conduct energy audits for the factories often do not prioritize energy efficiency for financial and technical reasons. The overall objective of this programme was to improve the energy efficiency levels in industries by promoting the practice of energy auditing. The programme's specific objectives were as follows:
- Promote energy auditing as an effective tool for industrial energy management;
- Building local energy auditors' capacity;

Table 6-5. Comparative Industry Sector Energy Performance Data

Industry Sub-Sector	Number of Sites	Typical Baseline Energy Consumption Per Site (toe/yr)	Total Baseline Energy Consumption Per Sub-Sector (toe/yr)
Food	208	519.98	108155.84
Iron & Steel	95	4002.37	380225.15
Rubber	145	1691.15	245216.75
Cement	5	12807.00	64035.00
Ceramic	23	1112.20	25580.60
Glass	14	5709.00	79926.00
Paper	48	2447.40	117475.20
Wood	203	19078.43	3872921.29

- Develop a sustainable energy audit programme in industries; and
- Maintain, calibrate and rent out audit equipment.

Based on the preliminary audits conducted by the MIEEIP team, it has been found that there is high potential for arresting energy wastage or loss by improving the design and construction of industrial facilities.

Energy Rating

It has been shown that minimum energy efficiency standards for equipment and machinery can be used as a cost-effective option to improve overall energy efficiency. The overall objective of this component has been to introduce activities that would inform industries about energy efficient equipment/machinery, particularly cost, technical specifications, economics and energy performance. The immediate objectives of this programme have been as follows:
- To provide information on energy efficient equipment and energy rating programmes;
- To increase awareness of, and encourage the use of, energy efficient equipment within the industry; and
- To set up an industrial testing facility to label equipment.

Energy Standards

Under the MIEEIP, a sub working group on motors was set up to promote the use of high efficiency motors. This group agreed to adopt the European Committee of Manufactures of EU Electrical Machinery and Power Electronics (European CEMEP) scheme as the basis for the Malaysian standard for motors. Based on an impact analysis, the team concluded that the introduction of the European CEMEP (Class 1 and Class 2) in two phases, could bring about potential savings of 72,000 GWh for the period between 2003 and 2004. In December of 2003, eight companies from the electric motor manufacturing, supply, and import industries signed a voluntary agreement to phase out inefficient motors by the year 2008. This agreement addressed energy labeling and industry motor replacement options.

Promotion

A main objective of this programme has been to disseminate information on energy efficient practices and technology applications to industries. This is being done specifically through the publication of a quarterly newsletter, MIEEIP News, and a well-equipped resource centre that caters to a diverse audience.

A significant activity under this programme was the establishment of an association of energy professionals, consultants, technology developers and providers. The Malaysian Energy Professionals Association (MEPA) was officially registered in 2002, and it serves as a platform for energy professionals to discuss matters on, and disseminate information related to, energy efficiency. To date, it has 150 registered members.

This programme has submitted to the Ministry of Energy, Water and Communications, a comprehensive proposal on the registration of energy service companies (ESCOs). This scheme will serve to ensure the credibility and professionalism of ESCOs in Malaysia once it comes into force.

ESCO Support

The rapid growth of the industrial sector has strained the capacity of the sector having to focus on multiple objectives. There is opportunity for ESCOs to assist the industrial sector in achieving energy efficiency objectives without compromising industrial productivity and growth.

The overall objective of this programme called for the successful development of an ESCO support programme in creating a business sector capable of meeting the energy efficiency challenges of the identified industrial sectors of Malaysia.

The following workshops, seminars and training sessions on business plans, financing, energy engineering & design tools and energy auditing to build the capacity of local ESCOs have been organized.
- ESCO Business Development Workshop
- ESCO Comprehensive Development Workshop
- Mini Workshop on Energy Performance Contract (EPC).
- National ESCO Workshop
- Financial Institutions Forum
- Mini Workshop on Model of ESCO Business Plan and Strategy
- Individual consultancy for seven ESCOs in their respective business
- Industrial Energy Performance Contract (EPC) Workshop
- Energy Engineering and Business Tools Workshop
- Energy Engineering Tools Workshop follow-up for ESCO
- Business and Financial Planning Workshop
- "Hands on" energy auditing training for eight ESCOs in the industrial sector and twelve ESCOs in the building sector.

The transfer of knowledge and experience, facilitated by both local and foreign experts, has assisted the ESCOs to prepare bankable proposals and market their services to a wider clientele. The programme has also achieved the following:

- A suitable institutional and legal framework for ESCO activities in the country has been developed;

- Institutional arrangements that promote ESCOs to the industrial sector are in place; and

- ESCOs have been aided in defining the feasible products and services that they can offer as well as to evaluate the risks associated with performance contracting.

What is an ESCO in Malaysia?

An ESCO, or energy service company, is one that is trained in providing energy management services to industries/clients. Most ESCOs began by providing engineering services or consulting to clients, but later, they moved into providing financial solutions as well as risk mitigations for energy efficiency activities. The ESCOs then began backing their services with performance guarantees, securing the financing for projects, and introducing a method of loan repayment from the savings enjoyed.

If customers answer 'Yes' to any or all of the questions below, they are most likely in need of assistance from an ESCO to implement energy efficiency improvement projects.

- Need help in identifying and implementing energy efficiency projects?
- Lack of available/experienced staff to install and manage the project?
- Lack of available/experienced staff to maintain equipment?
- Lack of financing?

Energy Performance Contracting (EPC)

One of the project goals is to promote an implementation scheme for energy efficiency projects that is more attractive than the normal engineering solutions provided by equipment engineers and suppliers. The MIEEIP has adopted the internationally recognized energy performance contracting (EPC) approach that will offer a more viable solution to the industries, especially in a financial market that has not included energy efficiency as a part of a potential portfolio.

The fundamental principle of the EPC approach is the implementation of energy efficiency projects by transferring the financial, technology and management risks away from the energy end-user to the ESCO.

Among the attributes of performance contracting are:

- Most of the technical, financial and operational risks are borne by the ESCO;

- EPC offers turnkey services, including feasibility analysis, design, engineering, installation and commissioning, maintenance and financing; and

- ESCOs are compensated based on the performance of the project.

The MESA

The Master Energy Services Agreement (MESA) was drawn up by MIEEIP team as a sample document to assist ESCOs and industries in the implementation of energy efficiency (EE) activities.

Demonstration Projects

In Malaysia, energy efficient technologies usually come with a high price tag, and industries are often reluctant to invest in activities that do not have visible benefits to productivity or profit levels. Besides, factories tend not to prioritize energy efficiency because of the relatively low price of energy.

To convince the industries of the importance of energy efficient technology applications, it is necessary to show them successful examples within similar working conditions. In addition to the actual physical working hardware, factories need to see evidence that such applications contribute to the improvement of the company's cash flow.

As such, the overall objective of this component was to demonstrate the applicability as well as the technical and economic feasibility of advanced energy efficient technologies in Malaysian industries. This was done by implementing one energy efficiency demonstration project in each of the eight industrial sectors.

The programme also:

- Disseminated information on the benefits of these technologies in the local industrial settings; and

- Provided technical and economic assistance to industrial energy users.

Seven companies were selected to participate in the MIEEIP demonstration projects; these factories were mainly from the 54 MIEEIP audited sites.

The demonstration projects each carried out two schemes, both aimed at capacity building for the local industries and the ESCOs. Under the normal approach, PTM worked with six of the seven companies as a joint venture. Whereas, the other scheme was implemented under the ESCO "fast track" approach. One demonstration project has been successfully implemented at Heveaboard Berhad, a company of the wood sector.

The financing of both approaches was partly from the MIEEIP's RM 16 million Energy Efficiency Project Lending Scheme (EEPLS) set up for the implementation of the demonstration project. The Malaysian Industrial Development Finance Berhad (MIDF) is managing the EEPLS.

For the normal approach, RM 8 million has been allocated at zero percent interest. Loans of up to 50 percent of the project cost or RM 2 million, whichever is lower, will be provided for the selected factories or host sites, which will have provide the balance of the investment.

Local Equipment Manufacturers Programme

Feedback from local industrial firms indicates that most of the locally manufactured industrial equipment is of poor quality when it comes to energy efficiency. But this equipment finds its way into the market because of the relatively high price of imported equipment. Unfortunately, in order to save on operating costs, industrial firms are burdened with equipment breakdowns and high energy bills because of the inefficiencies.

This programme aims to identify the potential improvements and new designs for locally manufactured industrial equipment. While the demand for energy efficient equipment is being addressed by the other components, this component will ensure that local manufacturers are motivated to improve their designs and manufacturing techniques. The programme has achieved this by:

- Training local equipment manufacturers on high efficiency designs and production technologies;

- Providing technical assistance to local equipment manufacturers;

- Providing funds to projects for design and manufacturing improvement of selected local industrial equipment manufacturers; and

- Evaluating the results and impacts of the funded equipment design and manufacturing improvement projects.

An equipment market survey identified five sets of equipment crucial in the efforts to improve industrial energy efficiency, which are locally manufactured. The following three types of equipment have been selected for design and manufacturing improvement under the MIEEIP.

- Pumps
- Fans & Blowers
- Motor Re-winders

EE Financing

The lack of participation from the local financial institutions in EE projects stems from their lack of awareness about the potential benefits of such ventures. Those who have been approached in the past by potential borrowers (ESCOs and industrial firms) were not convinced that the proposed projects were bankable or feasible. They view EE projects as high-risk and low-return ventures.

While the national energy policy provides a clear direction for energy-related activities in Malaysia, the absence of a comprehensive legal and regulatory framework is delaying the effective and sustainable implementation of EE projects. Specifically, the conventional financial loans are not viable for EE projects because of the government subsidy of energy prices, which is aimed at creating a business- and investment-friendly environment.

To remove these barriers, a number of efforts have been carried out, such as road shows on the need for a special scheme to cater to EE projects for selected financial institutions, and forums, to address the roles and opportunities for the financial sector in EE projects.

EE Project Loan Financing

It is generally assumed that industries would require external financial assistance when implementing EE activities. This is mainly because energy efficiency is not the core business of the factory or company, and the management would prioritize production over energy efficiency when deciding on investments.

For the MIEEIP, an EE project loan-financing scheme of RM 16 million has been setup at the Malaysian Industrial Development Finance Bhd (MIDF). As fund managers, MIDF has disbursed loans for several demonstration projects. The Global Environment Facilities (GEF) and the Malaysian Electric Supply Industry Trust Account (MESITA) make the allocations available.

MIDF has also agreed to be the fund manager for projects involving local manufacturers of energy efficient equipment.

Conclusion

The current status of the ESCO in Malaysia is not encouraging, despite the efforts by the government of Malaysia, initiatives like the MIEEIP and many other efforts. There should be more effort to emphasize, and act upon, the following issues:

- Follow through of the MIEEIP;
- Low electricity tariff;
- Attractive tax incentives;
- EE rules and regulations;
- Certification of ESCOs; and
- Certification of EE professionals.

The Malaysian government should be congratulated for recognizing the importance of energy efficiency and for taking an initial positive effort through the MIEEP.

VIETNAM

The concept of energy efficiency and conservation (EE&C) was first introduced to Vietnam in the 1990s as a part of technical and financial assistance programs provided by international organizations. A number of projects were implemented to study the use of energy in coal-fired thermal power generation, cement, ceramics, coal exploitation, metallurgy and other major industries, and to propose energy conservation measures applying to each sector.

As a result of nation-wide activities promoting EE&C in Vietnam, especially the Demand Side Management and Energy Efficiency (DSM/EE) project, a large number of engineering and trading companies have been involved in the new-born EE&C market. In collaboration with research institutions and energy conservation centers, these companies provide different types of services which can be categorized into:

- Equipment-based service: In order to sell the energy saving equipment; e.g., "thin" fluorescent tubes, variable speed/frequency drives

(VSD/VFD) and high-efficiency motors, an add-on energy audit service provided free-of-charge.

- Consultancy-based service: A large portion of profit comes from engineering work; e.g. energy audit and/or system modification. In most cases, new equipment was acquired by project owners.

- EPC-type service: Consultancy services and equipment acquisition will be provided in a package. The company can support the project owner on the financial arrangements; however, the project owner must take care of the loan.

Until 2004, most of financing sources for such EE&C projects came from international funding facilities (provided by international financial institutions such as the World Bank, or by developed governments) or government budgets. However, with the participation of small-and medium-sized enterprises (SMEs), as project owners in EE&C activities, the financing mechanism has changed. Many EE&C-related investment projects were financed partially by loans from commercial (private) banks. Recently, several projects were proposed to use partially, or solely, the client's own money or the money of the energy service provider.

Concerning contracting for EE&C services, most of projects were contracted so that the project owner will pay a certain amount of money for the consultancy services provided or equipment acquired. Payment based on the results of EE&C measures/recommendations can be high risk due to the following reasons:

- Insufficiency of legal guidelines/directives on supporting EE&C activities;

- The National Accounting System has not been adjusted to deal with characteristics of EE&C contracts based on guaranteed savings;

- Unavailability of financing sources; and

- The methods are not well defined and agreed between stakeholders.

However, with all efforts from the government of Vietnam and international organizations to create and maintain an open, competitiveness-based electricity market, it is expected that the first ESCOs will be established in Vietnam within three years.

INDIA

India is of increasing importance to the world economy, and energy is increasingly important to India's economic progress. Recognizing the vital role energy plays in all aspects of the Indian economy, a decade ago a Minister of Power for India wisely observed:

"No power is as costly as no power."

The country's ability to use its available energy with enhanced efficiency is a critical growth factor. In 2002, the Prime Minister called on the government to reduce energy consumption by 30 percent in the next five years and noted, "An effective method to achieve this target, which has already been proven in several public buildings, *is through contracts for guaranteed levels of energy efficiency improvements involving energy service companies....*"[emphasis supplied]

While the 30 percent goal was not realized, the energy service company model became recognized as an effective mechanism for delivering energy efficiency. Several donor agencies, including the World Bank, US AID, CIDA, DFID and GTZ, as well as Government of India (GOI) programs, encouraged the growth of the ESCO industry. In 2005, several ESCOs joined together to form an association, the Indian Council for Promotion of Energy Efficiency Business (ICPEEB). ICPEEB has aided in increasing the awareness of ESCOs and has worked to maintain dialogue with authorities for ESCO market development. The ESCO concept was piloted in nine states, averaging five to six projects per year from 2004 to 2007.

In a study conducted in 2007 by Econoler International for the Indian Renewable Energy Development Agency Ltd. (IREDA) and funded by the World Bank, 38 projects in the public sector developed by ESCOs were identified. These projects were being implemented by 16 ESCOs. The Econoler study revealed that they were typically small projects with 43 percent costing less that USD 300,000 and 57 percent requiring investment costs under USD 500,000.

Much of the information presented in this section is drawn from the IREDA study, *Analysis of Indian Experience with ESCO Delivery of Energy Efficiency for the Public Sector,* conducted by Econoler International and published in October 2007. As part of the study, the ESCO market potential in the public sector was analyzed. The results of this review are shown in Table 6-6.

Table 6-6. Indian ESCO Market Potential

Items	Unit	Public buildings	Street lighting	Total
Actual energy consumption	GWh/year	11,530	5,580	17,110
Total EE and RE technical potential	GWh/year	2,306	1,674	3,980
Max. investment potential	USD million	802	582	1,384
Technical and economical ESCO market	USD million	187	136	323
Technical and economical ESCO market	INR crores	861	625	1,486

In 2003, a study, *India Energy Outlook,* conducted for the Asian Development Bank estimated that India had an immediate energy savings potential of 5,300 million kWh and peak savings of 2,188 MW in the power sector alone, which corresponded to an investment of INR 17.2 billion. A summary of the findings is presented in Table 6-7.

According to the study, the investment size is estimated between USD 323 and USD 374 million. These conservative overviews indicate that the ESCO market in the Indian public sector is huge and most of it is still untapped.

Indian ESCO Activities Background

The development time line for India's ESCO industry effectively began in 1992 when an ESCO feasibility study was funded through USAID. Between 1992 and 1996, various programs funded by several donors led to feasibility studies, seminars, match-ups of select Indian companies with foreign ESCOs, and project implementation. During this period, Hansen

Table 6-7. Energy Efficiency Investment Potential in India

Market Type	Investment Potential (INR Billion)	Energy Savings (million kWh)	Energy Savings (MW)
Government (offices and hospitals)	4.2	1,600	500
Municipal	13.0	3,700	1,688
Total	17.2	5,300	2,188

Associates introduced the ESCO concept to many, provided bank training and conducted "match-ups" between Indian companies and US ESCOs. These efforts lead to further ESCO promotion and the creation of a few ESCOs. As an independent effort without donor participation, Hansen Associates also brought Thermax, in Pune, and a US ESCO, EPS, together to form an ESCO.

The years between 1999 and 2001 also saw the start-up of the ECO-I and ECO-II projects (which facilitated emergence of a few ESCOs and focused on best-practice DSM projects), and the extension of a USD 5 million GEF grant under a USD 130 million World Bank's credit line to IREDA. A portion of these funds were used to support ESCO market development and to build the capacity of ESCO stakeholders.

In 2001, the GOI passed the Energy Conservation Act and established a statutory body, the Bureau of Energy Efficiency (BEE), for regulation and promotion under the Ministry of Power. The BEE was officially established in March 2002. Its Action Plan was subsequently approved and released in August 2002.

The Econoler International study concluded:

Despite a decade of efforts to promote an EE market, the ESCO concept is still relatively unknown and unexplored in India. The absence of big players, the limited financial ability and inadequate past experience cause ESCOs to seem untrustworthy to clients as well as banks.

The public sector must provide leadership and guidance to EE markets by building confidence among stakeholders, demonstrating the financial viability of ESCO projects while tapping the energy savings potential in the sector. The central government made a commitment to reduce energy consumption by 30% in government's buildings and establishments. Steps were initiated to implement energy efficiency projects through the ESCO route in selected public buildings.

ESCO Activity in the Private Sector

A particularly active ESCO in the private sector in India at this time is DSCL Energy Services Company, Ltd. The energy service company is a subsidiary of DCM Shriram Consolidated Ltd. Group. According to Ms. Nisha Menon, Senior Consultant to DSCL Energy Services, on average,

DSCL has been taking three to five projects each year and completing one or two using the guaranteed savings model (supported with technical guarantees only). The track record of the firm's experience with the private sector is relatively better than the public sector because of delays within the public administration.

The development of the private sector market has been aided by the World Bank's Three Country Energy Efficiency Project, where a focus has been on getting energy efficiency financing for the private sector from commercial banks. In particular, the project has focused on three clusters; paper, steel and glass. However, to date none of the funds have gone to an ESCO. USAID's ECO Phase 3 program has a task which addresses financing in the context of small to medium sized enterprises (SME) plus ESCOs in India.

Ms. Menon reports, "The availability of money is not as much an issue as the terms on which it is available."

Barriers

A broad barrier in the private sector to the realization of an ESCO industry is the absence of information regarding the development of projects and management's lack of attention to identifying/conceptualizing energy service projects. The current purchasing model for ESCO services in India replicates the existing administrative rules and tendering process without any adjustment to the different nature of the ESCO business. This competitive bidding process is supposed to ensure that the client; i.e., the government, obtains services at the lowest cost, and also that the bidders have an opportunity to make their own assessment. However, the actual system has drawbacks and several ESCOs consider it as being so complicated and costly that they usually decide not to bid, thus the tendering process is often declared unfruitful.

The process is lengthy and contracts are sometimes not issued for months. This is a strong deterrent to ESCOs, especially for the small ESCOs.

Requests for proposals (RFPs) in the public sector usually include an audit report. The ESCO audit seldom concurs with the baseline information set in the owner's audit. The two parties are often in conflict. The owners fail to realize they are seeking guaranteed energy saving results from an ESCO, not just the financing of an audit. ESCOs are not banks. To the extent that the requested services are based on the owner's audit, the ESCO risks are increased. Further, the owner, often unnecessarily, pays for two audits.

In the public sector, contract decisions are typically made at the building owner's level and those responsible for the maintenance of the building and the utilities are frequently not involved. There is a lack of knowledge of the ESCO concept at the field staff level.

In noting that measurement and verification procedures (M&V) are another major barrier, the IREDA report indicated the following problems exist:

> The design and the implementation of a fair and easy to understand M&V protocol is required. The ESCOs are unable to develop such adequate protocols and the customers do not see this as crucial at the time of the contract signature, until the time comes for determining the energy savings. This can create potential conflicts between the parties and thus increase doubts about the ability of ESCOs to deliver the guaranteed results.

> Much is made of establishing the baseline and there is less emphasis on M&V in the ESCO business. Usually M&V is a straightforward engineering issue… Sometimes, ESCOs agree to adopt simpler methods that may be approximate, rather than a theoretically accurate one which is perceived as a considerable investment in measurement. M&V is a contractual issue and it is usually not a complex one. The responsibility is shared between the ESCOs and the customers. The lack of knowledge of most of the ESCOs can lead to some misunderstanding and approximate documentation. On the other side, the customers show difficulty in taking the cultural leap from making very few measurements in facilities to monitoring in detail all relevant influencing factors.

> The M&V relies also on the quality of the baseline preparation. The main barrier at that level is inadequate data, rather than the complexity of the issue. Baseline preparation takes time, as sufficient time must be allowed to consider the operating conditions that might prevail in the facility.

> Any variation from the initial baseline will have to be justified and decisions will need to be taken by the two parties for setting the baseline.

It was difficult to assess the real achievement of the projects implemented in public buildings as the basis of this evaluation is not properly set up. Bills may increase or remain stable after project implementation due to the impact of adjusting factors. The case of street lighting is different as a proven approach was already established and M&V development is less costly.

Despite the involvement of Leja Hattengadi, who served on the Policy Committee of the International Performance Measurement and Verification Protocol (IPMVP) and at the time was with Tata Consulting Engineers, and Satish Kumar, who served as the Technical Coordinator of IPMVP and is now serving as Chief of Party for IRG under a US AID program in India, there appears to be little recognition and acceptance of IPMVP by the Indian ESCOs.

Contracts and Financing

The contractual issues, especially how to make a fair, simple and usable contract for India, is among the difficulties met in the past. The ESCOs are suffering from the difficulty of having their customers fulfill their responsibilities, as any accountability is set at the client level. The interaction between ESCOs and civil servants does not appear to develop a win-win situation. ESCOs are seen as those who want to make money by potential public partners, who in turn do not consider the benefits of energy savings. The fact that many ESCO businesses are perceived as manufacturer, or vendor-based, doesn't seem to help in overcoming this mind set.

In India, only the shared savings model is strongly discussed. Many people who are familiar with the ESCO concept are not even aware that a guaranteed savings model exists, and that it could be beneficial to them, depending on the actual barriers to be mitigated in each case. Part of the concern about the type of contract offered is based on the belief that the only benefit that ESCOs bring to a customer is financing.

Contracting models in India appear to depend upon the respective project financing scheme. Currently, it is challenging for ESCOs to obtain bank credit and they usually utilize their own equity to finance an ESCO project in the public sector. The challenge is three-fold:
- Banks' perception that ESCOs creditworthiness should improve;
- Terms of lending and appraisal methodologies by banks need to be adjusted to the savings profile; and

- ESCOs need to capitalize themselves, and equip themselves, to provide performance guarantees and take higher investment risks.

To date, banks have focused on the borrower's assets. Large assets are usually owned by ESCOs, who are subsidiaries of manufacturing companies, but rarely by stand-alone ESCO businesses.

Typically, banks treat ESCO projects as normal commercial projects. The bank loan is based upon the creditworthiness of the company and is asset based. This is true even in the case of ICICI Bank, which has had a special credit line for ESCO projects.

ESCOs are requested to provide bank guarantees by public sector clients, and banks ask them to provide collateral.

Even though ESCOs are investing their own money and taking much of the risks, banks insist upon guarantees and security deposits. In the shared saving model, ESCOs are disbursing their own equity/loans during the project implementation and the client will not pay any Rupee before the savings are realized and the operational demonstration is done. As an example, the ESCOs will disburse at least 5-10 percent of the total investments for the detailed audit. As a consequence, should the ESCO drop the project, this initial investment would be a direct loss to the ESCO.

Even with a security deposit (escrow account), ESCOs don't feel comfortable with public payment procedures and payment delays. In the meantime, the ESCO must repay the bank loan installments as per schedule. For example, in one instance the local government failed to honor contract obligations related to work in an indoor municipal facility, and was 20 months behind the schedule of payments. The guaranteed saving model based on the payments by the client during the project implementation would obviously eliminate this payment security barrier.

Future ESCO Development

ESCO growth within the private sector has recently received support from the government. The BEE has recently received responses from nearly 50 companies and the agency is working to put in place a structure where ESCOs might only need to provide technical guarantees. Menon predicts that should this happen, activity in the private sector could finally take off.

A concerted effort to overcome the barriers identified above is needed if the ESCO industry is to fulfill its potential. At present there are twenty ESCOs in India. Most are serving the public sector market.

ESCO Projects in the Public Sector

The IREDA report, previously cited lists 38 ESCO projects implemented by 16 ESCOs. Table 6-8 summarizes these projects.

Table 6-8. Indian ESCO Projects in the Public Sector

Sub-sectors	Number	Comments
Projects identified	38	Two projects proposed by EE equipment vendors were not sanctioned by IREDA due to various issues. IREDA sanctioned 1 Project wherein the loan was not availed as the client implemented the project with own resources. 3 projects initiated by the Central Government in Delhi have not resulted in signing of ESCO contracts.
Public Buildings	16	Tendering ongoing for 7 projects
Hospitals	4	Tendering ongoing for 3 projects
Municipal Street Lighting	12	7 projects were implemented by the same ESCO
Biomass gasification projects	4	For fuel switching in Industries
Utility	1	Project sanctioned and financed by IREDA
Municipal Water Pumping	1	Cancelled after loan was sanctioned by the Bank; due to change in top management.

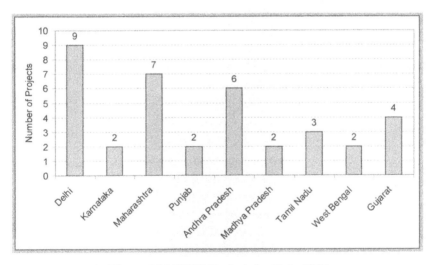

Figure 6-14. ESCO Projects by State, 2007

Out of the 38 projects identified in the public sector:

- 19 projects have been commissioned/contract concluded;
- 4 projects' contracts have been signed and the implementation was on-going at the time the data were collected;

- 9 ESCO projects for which RFPs have been issued or tendered; and
- 3 projects' contracts have not been signed.

Fifty percent of the projects noted in the IREDA report were located in New Delhi. Only one project used the guaranteed savings model; the others were all variations of the shared savings model.

With only 5 to 6 projects per year on average from 2004 to 2007, the trend is still slow and the IREDA report indicated that there has been no significant take off of the ESCO Industry in India.

CASE STUDY: Delhi Development Authority (DDA) Vikas Minar

Description of the project	The project consisted of implementing EE measures targeting the indoor lighting at the Delhi Development Authority buildings. The project was completed by Asian Electronics Limited (AEL) in April 2007. The actual annual energy savings achieved was 375,390 kWh per year against expected savings of 373,762 kWh per year. The payback calculated was five years.
Tendering process	• Initiated by DDA, BEE acted as advisor for the ESCO concept application. • The RFP was based on the CPWD documents prepared for their buildings. • Initial audit was done by consortium of DSCL, TERI, Petroleum Conservation Research Association (PCRA) and the RFP was based on the audit results. • Evaluation was delayed by two to three months. • There were five evaluation rounds before awarding the contract. During the interview, it was found that senior management and financial staffs had good knowledge of the ESCO process, but this was not the case for technical and field staff. The details regarding conducting of training workshops for dissemination of the ESCO concept are not known.
Contracting	• The shared saving models were adopted and copied from *BEE/CPWD* model. • Despite the ESCO taking all the risks and financing 100% of the project, DDA required:

- Bank guarantee for 10 percent of the total project cost for one year as a security deposit, until delivery and good installation of the project; and
- Bank guarantee for 10 percent of the payments for 5 years for the saving projects.
• DDA did not offer any payment guarantee through trust retention account or escrow. But DDA has, however, agreed to pay interest on delayed payments.

Financing	The total cost of the project was INR 8 million. It was funded by the ESCO equity and loan obtained by the ESCO from the Industrial Development Bank of India Limited (IDBI). There was no difficulty to get the loan from the bank, which accepted the receivables based on the existing track record of the ESCO.
Implementation and M&V	The project planning, procurement and implementation was completed within 5 months from January to May 2007 by the ESCO's own team. The measurement and verification plan adopted was the partially measured, retrofit isolation method done jointly by ESCO and DDA.
Project Replication	Three other projects in public buildings were awarded to AEL by CPWD and DDA and were recorded during the survey: Shram Shakti and Transport Bhawan, Prime Minister's Office and DDA Vikas Sadan.
Main Issues	• The audit report was used as basis for the RFP. The baseline consumption of the connected load/lamp was lower than the actual figures and ESCO was blamed for this as the savings provided in the proposal were lower.
• The evaluation process went through technological aspects not the savings part and even included comparison of prices of equipment in the market. Thus the services provided by the ESCO for maintenance and the financing costs were not considered in the evaluation.
• Payments were delayed because a window of 12 months was allowed *whereas* the construction took only 5 months. The same difficulty was faced for the whole project payment procedure.

PHILIPPINES

Energy efficiency and conservation in the Philippines at the present time, has been picking up where it left off during the past energy crises. Today, when oil prices are again soaring, the Philippine government is starting to re-focus its efforts on conserving energy. Ms. Alice Herrera provides valuable insight as to the current status, barriers and potential remedies that the ESCO industry faces in the Philippines.

The Philippine government officially launched the National Energy Efficiency and Conservation Program (NEECP) in August of 2004, as an essential strategy in rationalizing the country's demand for petroleum products and eventually lessening the impact of escalating prices on the economy. NEECP aims to provide the framework for the government's efforts to promote efficient and judicious utilization of energy. It is being popularized through a campaign dubbed as *"EC Way of Life."*

The NEECP estimated an aggregate energy savings of 24.5 Mtoe in 2014, which would come primarily from the intensified energy utilization management programs of the commercial and industrial sectors, power plants, utilities, as well as the continuous use of alternative fuels and technology.

One of the key strategies to increase the participation of the private sector in energy efficiency and conservation is to revitalize the ESCO industry.

ESCO Industry in the Philippines

The establishment of ESCOs in the Philippines started in the 1990s with the introduction of demand-side management (DSM) programs. An ESCO was then defined as a company which offered electro-mechanical design for increasing the efficiency of facilities and utilities, and provided energy audits, installed energy conservation devices, or monitored control of the electrical system and mechanical sewerage. At present, only a few ESCOs in the Philippines are doing general energy audits.

To better promote ESCOs, an association was established. In May 2005, the ESCOPhil was registered with the Securities and Exchange Commission (SEC) to organize the firms engaged in the energy service industry, with the main purpose of:

a) Providing a forum for the effective exchange of information about industry trends and practices, and for the introduction and propagation of new technologies for the industry;

b) Promoting energy efficiency and demand reduction technologies, thereby creating tangible economic values;

c) Developing strategic advocacy positions with government agencies;

d) Initiating policies geared towards increasing business opportunities for members; and

e) Educating and accrediting other firms and organizations as members.

At present, there is an on-going UNDP-GEF funded 5-year project, which started in 2005, to promote ESCOs focused on energy efficient lighting systems. This project is being implemented by the Philippine Department of Energy (DOE) and addresses the barriers to widespread utilization of energy efficient lighting systems (EELs) in the Philippines.

The strategies of the project include updating of policies, standards/ guidelines; institutional capacity building; consumer education and information dissemination; developing and implementing financing mechanisms; and mitigating environmental impacts. The implementation of the project will result in a projected aggregate energy savings of 29,000 GWh. This will be equivalent to 21 percent reduction relative to the Philippines energy efficiency scenario from 2003 to 2012.

Under the PELMATP, a model ESCO Transaction Project was implemented at the Development Bank of the Philippines (DBP) and an ESCO Demonstration Project at Banco de Oro (BDO) branches. This strategy aims to generate interest in ESCO projects within the commercial, industrial and institutional sectors, thereby contributing to energy consumption reduction and environmental protection.

Types of ESCO Contracts

ESCOs generally act as project developers for a wide range of tasks and responsibilities and assume the technical and performance risk associated with the project. Typically, they offer the following services: develop, design, and finance energy efficiency projects; install and maintain the energy efficient equipment involved; measure, monitor, and verify the project's energy savings; and assume the risk that the project will save the amount of projected energy savings.

There are a number of ways to contract with ESCOs depending on the degree of risk that the company will assume. If the ESCO assumes all the risks and guarantees project performance, the cost for their services will be higher. There are three main types of ESCO contracts, the shared savings, the guaranteed savings and the fee for service contracts. In the Philippines, companies still prefer the "fee for service" type of ESCOs as seen in Figure 6-15.

Under a *guaranteed savings* contract, the ESCO guarantees a given value of energy savings. The customer makes periodic debt service payments to pay off the cost paid to the ESCO for developing, designing and installing the efficiency measures. If the guaranteed savings level is not achieved, the ESCO covers the difference between the guaranteed savings and the actual savings. However, the client keeps any savings above and beyond the guaranteed savings level unless stated otherwise.

Under a *shared savings* contract, the customer commits to pay only a percentage of the realized cost savings to the ESCO for the costs of designing, implementing and monitoring the energy efficiency project. In this type of contract, the customer assumes no financial obligation other than to pay the ESCO a share of the savings that the project realizes. Thus, the ESCO, which finances the project assumes both project performance risk and the credit risk. Since shared savings is vulnerable to energy price volatility, this approach is more suitable for projects with short payback periods.

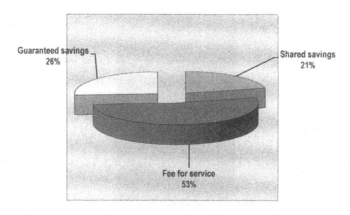

Figure 6-15. Types of ESCO Contracts in the Philippines.
(Source: Nezhad, H.G., 2004)

ESCOs are currently experiencing difficulties setting up their business because of some financial and institutional issues. ESCOs have difficulty accessing local debt financing because of "credibility" issues; lack of information about energy performance contracting (EPC) among financial institutions and the requirement for assets as collateral. Local financial institutions require real estate as collateral for loans which the ESCOs cannot provide as they are not asset-based, but service-based. No incentives are being given by the government to ESCOs or to EE projects, in general.

ESCO Gaps, Constraints and Barriers

In 2004, the Asian Development Bank (ADB) conducted an online ESCO survey, which aimed at determining the past and current activities of ESCOs. The survey also focused on the identification of the critical success and risk factors of ESCOs, as well as the problems and opportunities for energy efficiency improvements and ESCO development. According to the survey, the critical success factor was determined by the financial capability of the ESCO, while cash flow was the most important risk factor.

The ESCO Specialist under the UNDP-GEF-DOE PELMAT project has identified four (4) major barriers as follows:

1. Technical Barriers:
    - Absence of ESCO accreditation for energy audit services and/or EPC; and
    - Few ESCOs provide general energy audit services.

2. Financial Barriers:
    - Absence of financing programs to industry and/or ESCOs; and
    - No available government incentives.

3. Institutional Barriers:
    - Weak EPC concept propagation to industry and financing institutions;
    - "Credibility" issue due to previous industry experience;
    - Young ESCO organizations; and
    - No energy management or energy conservation center.

4. Regulatory Barriers:
    - No applicable energy conservation law; and

- No prevailing ESCO ownership structure for local investors/ players.

Proposed ESCO Framework of Cooperation

Under the UNDP/DOE PELMAT project, a model ESCO transaction project was prepared with the Development Bank of the Philippines (DBP) and Banco de Oro (BDO) as cooperating facilities. DBP is a state-owned national development bank while BDO is a private financial institution. The project provided an ESCO specialist to assist these financial institutions in fully understanding the concept of energy performance contracting. It also involved capacity building for commercial and industrial establishments on ESCO operations, financing, and development of measurement and verification procedures. The ESCO specialist provided assistance to DBP in the development of request for proposals (RFP) and bid documents, drafting of ESCO documents, such as energy service agreements (ESA), ESCO financing process flow, ESCO Project Financing Program, ESCO Term Loan Agreement, and ESCO project evaluation and endorsement report. Assistance was also extended in preparing justification position papers to amend some of the procurement policies of the government to accommodate the ESCO concept.

Three shared savings frameworks of cooperation were presented and these were from Trigen-Energy Philippines, Inc., OSP Honeywell and KC Industrial. Figures 6-16 and 6-17 for Trigen and Honeywell respectively show the use of third party financing. Figure 6-18 presents the framework using ESCO financing.

The Way Forward

The following are some of the recommendations to further promote ESCOs in the different sectors of economy in the Philippines:

- Develop and institutionalize an ESCO accreditation scheme. This includes defining and categorizing different types of ESCOs.

- Strengthen information dissemination and capacity building on ESCOs and energy performance contracting schemes among the industrial and commercial sectors.

- Develop different financing portfolio schemes for ESCOs and provide fiscal and non-fiscal incentives for ESCO projects.

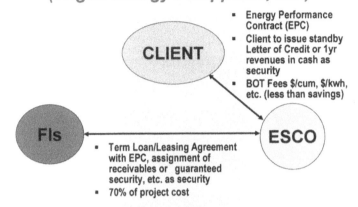

Figure 6-16. ESCO Framework of Cooperation (Trigen-Energy)

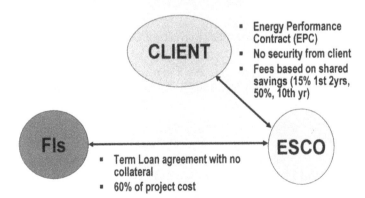

Figure 6-17. ESCO Framework of Cooperation (OSP Honeywell)

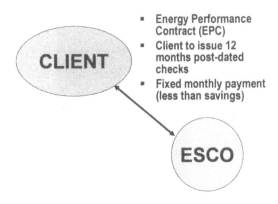

Source: Raymond A. Marquez, ESCO Specialist

Figure 6-18. ESCO Framework of Cooperation (KC Industrial)

REPUBLIC OF KOREA

Mr. Ed Sugay of Siemens Building Technology provides a comprehensive look at the current status of the ESCO industry in the Republic of Korea (South Korea).

The Korean ESCO business in the building and commercial sectors is in its nascent stage.

Although the government has been proactive in developing policy and promulgating legislation with the goals of ensuring energy security and promoting energy efficiency—for example, the creation of the Ministry of Energy and Resources in 1977 and the subsequent promulgation of the Rational Energy Utilization Act (REUA) in 1979. The Korean energy services business has not seen the rapid rates of growth over the past 2 ½ decades compared to its counterpart in the major developed countries in Europe and North America.

The Korean energy services companies have reported performing an increasing number of energy efficiency projects in the heavy industrial sector (steel, cement, semiconductor, manufacturing, chemical and refining, etc.). However, considering the very high population density and the large number of older high-rise commercial buildings in Seoul City and its surrounding counties, the relatively high cost of energy and the need for winter heating and summer cooling, the energy services business in the building and commercial sectors has not experienced a robust and sustainable growth trend in the past several years.

Brief Summary of the Energy Situation in Korea

The Republic of Korea, located in southern half of the Korean Peninsula, is one of the four Asian Tigers that experienced tremendous economic growth from the 1980s to 1997. In the 1960s, its gross domestic product (GDP) per capita was low and comparable to levels in poorer Asian and African countries. By 1995, after more than a decade of GDP growth in the 7 percent to 10 percent range, Korean GDP per capita surpassed that of some European Union (EU) countries.

The Asian financial crisis that started in Thailand in 1997 affected Korea in a very profound way. For example, due to the rapid devaluation of the Korean Won and the need the Bank of Korea (central bank) had for increased dollar reserves, thousands of Koreans responded by donating their gold, jewelry and other assets to the treasury. Between 1998 and 2000, the Korean economy contracted (negative GDP growth rates of -7 percent, -6.5 percent and -4 percent, respectively), but has since rebounded. GDP growth rates in the years from 2001 to 2006 varied from 3.1 percent to 7.0 percent. (Source: Asian Development Bank.)

The Korean economy of today is dominated by manufacturing, particularly of semiconductors and electronic goods, passenger vehicles, heavy machinery and chemicals. Agriculture, forestry and fisheries, which made up 25 percent of total GDP in 1973, had fallen to less than 3 percent by 2005.

Korea has very limited supplies of indigenous natural resources. It has no domestic oil resources and only a very small amount of natural gas has been produced locally (starting in 2005). Given Korea's exposure to rising world commodity prices, the government placed the cost of energy imports in 2005 at USD 66 billion, a 33 percent increase from 2004.

Consumption of natural gas in the industrial sector has grown eleven fold in the last decade—equal to an average annual increase of 27 percent. Oil accounts for a relatively large share of industrial consumption at 57 percent of the total in 2004. Consumption in the transport sector reached 24 percent in 2004. Outside the transport and industrial sectors, energy use increased by 28 percent. In particular, residential energy consumption has more than doubled since 1995.

With very limited indigenous energy resources, Korea is particularly concerned with security of supply and energy diversification. The government has focused on the increased use of natural gas, bituminous coal and nuclear power. It is also aiming to reduce its reliance on imported oil, as well as to diversify its import sources in order to reduce risks from volume and price fluctuations of crude oil.

Today, Korea's overall energy policy seeks to achieve sustainable development through energy security, energy efficiency and environmental protection (the 3 E's policy). This is a marked departure from its historical energy policy of sustained economic growth to maintain a high-quality of life for Korean citizens. Korea's energy policy now promotes stable energy supply, market efficiency through competition and the implementation of an environment-friendly energy system with the end-goal of sustainable development.

Business Drivers for the Energy Service Market
Government Policies and Incentives

In 2008, the Korean government created the Ministry of Knowledge Economy (MKE) to replace the Ministry of Commerce, Industry and Energy (MOCIE). In addition to the former functions of the MOCIE, the new MKE also incorporates other functions that were previously the responsibility of other ministries; e.g., information and communications, science and technology, finance and economy. The mission of the MKE is to develop Korea into a knowledge-based economy, one that is driven by technological innovation. Korea's energy conservation and environmental initiatives and programs which are planned based on the REUA, will be implemented by the MKE through the Korea Energy Management Corporation (KEMCO). KEMCO, which was established in 1980, based on the REUA, is the government agency in charge of implementing the national energy efficiency and conservation policies and programs.

KEMCO's major activities are:
- Implementing energy-saving programs by sector of energy use;
- Conducting energy audits and surveys;
- Promoting energy efficiency;
- Commercializing and diffusing the use of higher-efficiency energy appliances;
- Conducting R&D to demonstrate and disseminate technologies on energy and mineral resources; and

- Leading the effort to mitigate climate change.

The following subsections are government initiatives and programs that KEMCO is implementing.
- **Voluntary Agreement.** The Voluntary Agreement (VA) is a joint program between the government and industry that is managed by MOCIE and the Ministry of Environment. A company which intends to join the agreement submits a concrete action plan that specifies its energy consumption and GHG emission reduction targets. Upon approval of the action plan by KEMCO, the qualified company and KEMCO execute the VA contract. The qualified company is then supported with low interest loans and tax incentives for energy conservation and GHG reduction, as well as technological support to achieve its goals.

- Between 1999 and 2004, there were 1,021 companies that signed VA contracts, and KEMCO reports that the total investment for this period was USD 3.86 billion.

- **Preferential Loans.** The government provides long-term low interest loans with rates between 2.25-3.5 percent with three to five-year grace periods and five-year repayment periods, for installation of energy saving facilities or equipment. Up to 100 percent of the investment amount can be provided to applicants. The maximum loan amounts per project range from USD 10 million to USD 25 million.

    Other forms of government support for ESCOs include: credit loans, mortgage programs and working capital loans. In 2006, the government funded the Rational Energy Utilization Fund with USD 655 million, which KEMCO disbursed as incentives to ESCOs and SME (small- and medium-size enterprises), through direct loans, loan guarantees and working capital loans.

    Income tax credits are available for: (1) replacement of old industrial kilns; (2) installation of energy-saving facilities or equipment; (3) alternative fuel facilities; and (4) other facilities that are judged to reduce energy consumption by more than 10 percent. The income tax credit was 10 percent in 2001-02, and 7 percent in 2003-06.

- **Energy Audits.** Energy audits in Korea have been conducted mainly by KEMCO. In the industrial sector, KEMCO offers two kinds of energy audits: an in-depth audit and a free audit. The in-depth audit (also called technical service audit) is conducted at the request of the user, while the free audit is provided to small, and medium-sized enterprises (SME), whose annual energy consumption is between 800 to 20,000 MWh.

- In the commercial building sector, in-depth energy audits are conducted for large commercial and residential buildings at the request of the owners of the buildings. Depending on the results of the audits, KEMCO will provide technical assistance and energy efficiency improvements, such as thermal insulation and double-glazed windows. In 2004, KEMCO reported the following data for in-depth audits that it had conducted.

High Energy Costs

As shown in Table 6-10 next to Japan, Korea has the highest energy costs in Asia. In 2005, Korea's electricity end-use prices, including taxes and converted using current exchange rates, are higher than that in the United States.

Installed capacity by fuel type at the end of 2006, per Korea Electric Power Company (KEPCO) is shown in Figure 6-19.

High Levels of Pollution

The Organization of Economic Cooperation and Development (OECD) released its Environmental Performance Review of Korea at the end of 2006. The report said that Korea ranked number one among the 30 member states in terms of energy intensity (energy use per unit of GDP). At 0.23 TOE (tons of oil equivalent) per USD 1,000 of GDP, Korea's energy intensity is well above the OECD average of 0.19, much higher than that of France (0.17) or Japan (0.15), and even higher than that of the USA (0.22).

Table 6-9. Potential Cost Savings through KEMCO In-Depth Audits, 2004

	Energy Cost Savings (USD million/yr)	Investment Cost (USD million)	Payback Period (Years)
Industrial Sector	129.5	242.0	1.9
Building Sector	0.89	1.997	2.2

Table 6-10. Relative Electricity Prices
Source: USEIA Monthly Energy Review, May 2007, Table 9.9

Electricity Prices for Industry (USD per kWh)[1]			
	2003	2004	2005
Korea	0.051	0.053	0.059
Japan	0.122	0.127	0.121
United States	0.051	0.053	0.057

Source: USEIA Monthly Energy Review, May 2007, Table 9.9

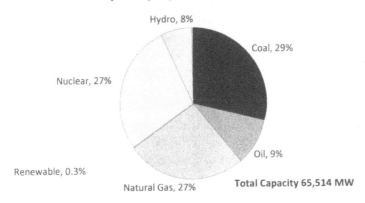

Figure 6-19. Korean Installed Capacity by Fuel Type, 2006

In GHG emissions, Korea's CO_2 emissions relative to GDP are among the highest of OECD countries and double those of France. In 2006, Korea's CO_2 emissions intensity, at 0.51 tons per USD 1,000 GDP, is just behind that of the US (0.55), significantly above the average for OECD countries (0.45) and that of Japan (0.36), and double that of France (0.24).

The OECD report made a total of 54 recommendations in the broad areas of energy and environmental management, sustainable growth and international cooperation. While the recommendations are not binding, Korean government officials said they will be considered when new policies are implemented.

Clinton Climate Change Initiative

In 2006, former US President Bill Clinton launched the Clinton Climate Initiative (CCI) of the William J. Clinton Foundation, with the stated "mission of applying the Foundation's business-oriented approach to the fight against climate change in practical, measurable and significant ways."

During the C40 Large Cities Summit in 2007, President Clinton announced the launch of the Energy Efficiency Building Retrofit Program (EEBRP), the first major program of the CCI. At its launch, the EEBRP program brought together four of the world's largest energy service companies (Siemens, Honeywell, Johnson Controls and Trane), five of the world's largest financial institutions (ABN AMRO, Citi, Deutsche Bank, JP Morgan Chase and UBS), and fifteen of the world's largest cities—including Seoul—in an effort to reduce energy consumption in existing buildings across the municipal, private, commercial, educational and public housing sectors. Existing buildings in the largest urban areas are the main focus of the program because buildings account for up to 70 percent of GHG emissions in cities such as New York, Shanghai and Seoul.

The EEBRP program began in Seoul in early 2008, with the appointment of the Seoul "CCI City Director," who is responsible for coordinating and facilitating cooperative activities among the various stakeholders, e.g., owners of Seoul City's building stock (municipal, commercial, private and educational buildings), financial institutions and the ESCOs. At the time of this writing, Seoul City was in the process of releasing a Request for Proposals for a pool of ten municipal buildings.

The Evolution and Current Energy Service Market

The energy services program in Korea started in 1992 with three registered companies. By May 2008, there were 160 ESCO companies registered with KEMCO. The major focus of ESCO solutions/services are:
- High efficiency lighting
- Waste heat recovery
- Heating and cooling systems
- Process improvements
- CHP (combined heat and power)

The following diagrams, Figures 20-22, show the evolution and growth of the ESCO market in Korea, and levels of investments, by ESCO solution/service and by business sector as reported by KEMCO.

Asia

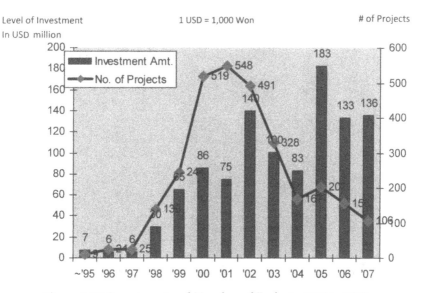

Figure 6-20. Investment and Number of Projects, 1995 to 2007.
(Source: KEMCO)

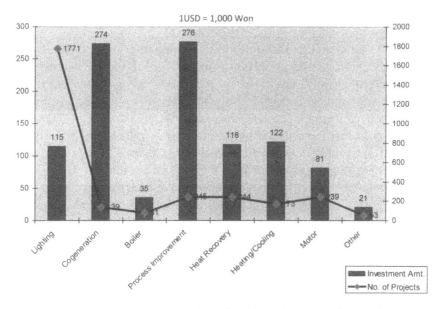

Figure 6-21. Investment and Number of Projects by Type of Retrofit, 1995 to 2007. *(Source: KEMCO)*

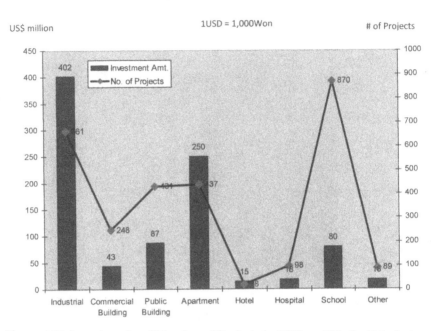

Figure 6-22. Investment and Number of Projects in Different Vertical Market 1995 to 2006

KAESCO—Korean Association of ESCO Companies

The KAESCO was established in 1999. The KAESCO website (in Korean language only) reports, as of the date of this writing, that it has 64 member companies. KAESCO's stated goals are to provide a forum of "friendship," information sharing and cooperation among ESCO companies and customers, and to positively impact the environment by facilitating the growth of the energy efficiency market.

KAESCO funding sources are:
- Membership fees (nominal entrance and annual fees);
- Receipt of a percentage of each project sold by its member companies (for example, for project sizes between USD 100,000 and USD 500,000, the fee is USD 450; for USD 1 million project, the fee is USD 1,500), and
- Special assessment as decided by the KAESCO "Directors' committee" (similar to a governing board).

KAESCO publishes a Korean ESCO magazine on a bi-monthly basis, funded by vendor advertisements. At the current time, KAESCO does not organize seminars or conferences and they do not provide training courses.

Financing Model

The KEMCO website describes the ESCO function as follows: "ESCOs invest in energy utilizing facilities on behalf of the energy user, if the user is unable to replace or improve existing facilities with ones more energy efficient due to technical or financial problems. *ESCOs bear the entire financial burden necessary for facility investment* and provide professional services for energy conservation facilities. The reimbursement of ESCOs' investment is made by saved money (*sic*) through future energy cost reductions." (From http://www.kemco.or.kr/english/sub03_esco.asp as of April 30, 2008; emphasis by authors.) This describes the "shared savings model" where the ESCO is asked to take the customer's credit risk ("if the user is unable to replace or improve existing facilities... due to financial problems"), a risk that arguably the ESCO is not most qualified to take.

KEMCO is aware of this challenge and concedes that the "guaranteed savings model" (in which the ESCO takes the *technical* risk by providing a guarantee of the technical performance of its installation work while shifting the customer credit risk to a third party is also needed to spur faster growth of the industry. To this end, KEMCO stated in earlier meetings that in fiscal 2008, a substantial portion of the Preferential Loan Program budget will be allocated to projects that use the guaranteed savings model.

Lack of Suitable Financial Intermediation

The Korean financial industry is quite mature; financial products and services offered are comparable to those offered in Hong Kong and Singapore. However, there are no local funding sources that finance ESCO projects because the model is new and local banks do not have windows for the more complicated structured financial products. Consequently, Korean ESCOs have been using internal capital to finance customer projects, thereby taking customer credit risk. This practice of taking customer credit risk and the widespread use of the "shared savings model" also explains the accounts receivables problems reported by some of the multinational ESCOs.

KEMCO has reportedly started an education campaign to raise the level of local banks' expertise in technical analysis and risk management for ESCO project financing.

CHINA

We are privileged to have views of the ESCO development in the Peoples Republic of China provided by Shen LongHai, the executive director and Zhao Ming, (Lily), secretary general of the EMCA, the Chinese ESCO association. They provide a good overview of the steps and leadership offered to move energy conservation, and ESCOs, from a planned economy toward a market economy in a relatively short time.

In the transition period from a planning economy to a market economy, the existing system for energy conservation could not properly meet the market demand and requirements, even becoming a kind of barrier to the promotion of energy conservation in China. It is a major task to achieve a market transition and to switch the energy conservation mechanism in order to ensure the healthy and sustainable development of the nation and the whole society.

In the planning period, the energy conservation works were carried out by a national plan, when the national government set up the target and plan for local provinces and for each state-owned enterprise. In all the sectors related to energy conservation, energy efficiency (EE) projects were implemented by administrative measures, from organization, technology, and investment. With the withdrawal of the planned economy and the development of a market economy, the investment and the operation behaviors of enterprises are becoming centered on economic benefits. The past administrative measures do not work effectively any longer. Therefore, building up of market-based mechanisms of energy conservation, and forming the energy conservation service industry, has become the most urgent need of the nation.

Introduction of Energy Performance Contracting

In 1998, the World Bank/Global Environmental Facility (WB/GEF) started the China Energy Conservation Project, cooperating with the Chinese Government on energy conservation and emission reduction. It is a very big measure to switch the EE mechanism from planned to market economy. The purpose of the project was to introduce, demonstrate and promote the energy performance contracting (EPC) mechanism, enhance the mechanism's familiarity, and increase energy efficiency by implementing energy conservation projects in China, thereby reducing emissions and protecting the global environment.

Phase I of the Project

Phase I of the project started in late 1998 with the main objective of setting up three demonstration energy management companies (EMCos) in Beijing, Shandong and Liaoning respectively, and establishing the national energy conservation information dissemination center. Following intensive training and consultation by Hansen Associates, funded by GEF, the success of the three pilot companies proved that EPC has great development potential in China and it is feasible and profitable to have more EMCos in China.

By the end of June 2006, the three pilot EMCos had implemented 475 EPC projects for 405 customers. The accumulated investments had reached USD 190 million, and the energy conservation capacity is about 1.49 million tce/year, CO_2 emission reduction capacity 1.45 million tc/year.

Phase II of the Project

Phase II of the project was officially launched in 2003, with the set target of further promoting EPC in the whole country, enhancing the development of EMCos (ESCOs) and forming the EMCo industry.

Detailed tasks for Phase II of the project included technical assistance for the new and potential EMCos to help them build the necessary capacity for implementing EPC projects and launching the EMCo Loan Guarantee Program, which provide financial support for EMCos.

Achievements of the Project

With the support of the project, a national industry association was set up—China Energy Management Company Association (EMCA). EMCA is a registered NGO with the mission of "Promoting the sustainable development of the energy conservation service industry, supporting the fast and healthy growth of EMCos" in China. The objectives of EMCA are to promote EPC and the development of EMCos by providing technical assistance through various measures, then forming and developing the China energy conservation service industry.

By the end of 2007, EMCA registered membership reached 300 from 59 in 2004, the investment of EPC projects reached about USD 1 billion. Through the EPC loan guarantee program, the Investment Guarantee Company has supported 36 EMCos to get commercial loans for 113 EPC projects, with a total investment of USD 83.85 million, and the total guarantee of about USD 49.43 million.

Development of ESCOs

Promoted by the WB/GEF project, and in the critical transition period of the Chinese economy, the EPC mechanism was successfully introduced. The EPC mechanism can play an active role in promoting energy conservation in China's market economy. This enhances the belief and confidence to further develop the energy conservation service industry in China while attracting more and more people to get involved.

The EPC mechanism has been spread widely and has been further promoted. The industry is developing, the capacity of the EMCos is improving, the financing channel is widening, and the best time for the industry development has come.

The Chinese energy conservation service industry has grown from nothing to prosperous, from small to big, from weak to strong. A basic team working in construction, heavy industry and transportation has been formed. Different types of EPC projects have been implemented. Now the EMCo industry has become an important player in the energy conservation field in China.

After several years of development, ESCOs in China have been developing the EPC models in several ways, based on the basic concept of shared savings. At present mainly there are three types of EPC contracts in China.

Types of EPC Contracts

- *Shared savings*: EMCo provides investment and services, while the host enterprise cooperates with the EMCo in the project implementation. EMCo shares energy savings benefits according to agreed proportions within the contract terms. After the contract has expired, the host enterprise retains all equipment and future energy saving benefits.

- *Savings guarantee*: Host enterprise can co-finance the EE project with the ESCO. The ESCO provides the turn-key services and guaranteed energy saving result. According to the contract, host enterprises pay service fees to ESCO/EMCo, and if the promised energy savings are not achieved, EMCo assumes corresponding responsibilities and compensate for the missing part.

- *Outsourcing of the energy management system*: Host enterprise provides promissory advance charge for EMCo operating or conducting renovation on the excising energy system. EMCo reduces energy expense (deducting new added management fee) through improving energy efficiency, and retains all or partial energy costs saved, according to the contract.

Based on these three types of basic EPC contracts, a diversity of complex contracts can be formed according to the practical situation of each project.

Author's note: The EPC contracts described above do not fit the shared savings models defined in Chapter 1 and used in most of this book. For example, client payments under the Chinese shared savings scheme are stipulated and the EMCo does not incur a performance risk.

Types of ESCOs in China

Using an EPC mechanism to implement EE projects is the common characteristic of EMCos in China. However, due to the differences of their key resources, the foundation of each ESCO and the major advantages of their operation differ. The following is a brief description of the kinds of ESCOs, which are active in China:

- *Reliance on technology and products*: These EMCos are based on energy conservation technology and products. Their core competitiveness has been formed while establishing the company. They seek new financing channels to obtain bigger market share. Most of these types of EMCos own their own intellectual property. They have the controllable technical risks in project implementation, higher project benefits and clear orientation for a target market, which is favorable for forming the competitiveness in certain industries.

- *Reliance on capital*: With these three pilot EMCos as the representatives, including a few new EMCos, their distinct advantage is abundant capital, and their business characteristic is to use their capital advantage to integrate energy conservation technology and energy conservation products to implement EE projects. This type of EMCo has more flexibility with more than one market and has the ability to implement EE projects in many industries with many technologies. However, this type of ESCO needs to strengthen the risk control capacity in selecting energy conservation technology, products, and in the project operation process.

- *Reliance on market*: This type of EMCo takes advantage of customer resources in certain industries. They use their controlled customer resource to implement corresponding energy conservation service projects. The cost of developing the market is comparatively low, and since they already know their customers, the customer risk is smaller. This favors long-term cooperative relationships and direct financing is more easily obtained for customers for EE projects.

Distribution of ESCOs and Their Service Fields

Growing out of the three pilot companies, there are now EMCos across China. The distribution is as follows:

- Of the 308 EMCos across China, most are in the North and East of China, with 45 percent and 18 percent respectively. There are fewer EMCos in Northeast and Northwest, with only 5 percent and 6 percent respectively.

- Service fields of ESCOs in China are three dominant market sectors: industry, building and traffic. Most of the EPC investment goes into the industry field with most projects implemented in the retrofit of the buildings.

Unlike the ESCO business in the USA, most of the ESCOs in China started their business in the heavy industry field, where the energy consumption is greater and the energy efficiency is lower. This provides more opportunities for business and higher return on investments. In recent years, the building field has become quite popular due to the large number of buildings, including universities, hospitals, commercial buildings, hotels and government buildings.

Opportunities

China is a big energy consuming nation, the primary energy consumption ranks second in the world. For a long period of time, China's economic development model has created great energy pressures and serious environmental problems, which have impacted on the social image of the nation and become a bottleneck for further economic and sustainable development.

Since the beginning of the new century, the Chinese government has paid ever greater attention to energy conservation and environmental protection. Numbers of relevant laws, regulations and national policies on energy conservation have been promulgated and published. The people's congress and the central government have set up strategic energy conservation goals; i.e., the energy consumption per GDP will be reduced about 20 percent and the main pollutant emission 10 percent.

Estimates show that there is great potential for energy conservation. The projects that are technically feasible and economically reasonable will exceed 500 million tc, with the market size to about USD 150 billion. That is a great market for the China energy service industry.

To ensure the successful completion of the 11th five year plan, the central government has published a series of encouraging policies and measures, such as implementation plans, etc. It is evident that the government is quite firm on pushing forward with energy conservation work. This creates a very favorable environment for the development of China's energy service industry.

Reducing energy consumption and GHG emission to protect the earth has become a common goal of the whole world. A trend has developed, which promotes the development of energy conservation and the energy service industry. The World Bank group and the GEF have provided great support to China financially and technically, which has prompted the fast development of China's energy service industry. The CDM mechanism (Kyoto Protocol) and the achievements from the recently closed-Bali conference also enhance the additional support from developed countries to the developing countries by providing energy efficient technologies and investment. This international environment will provide new opportunities for China.

China's energy conservation service industry is a quite new and small industry compared with already matured industries. But it is a good time to merge into the international environment and to shoulder the social and time mission, and will play a more and more important role in promoting social and economic development.

Major Barriers

New opportunities always go with great challenges. The more the opportunities are; the greater the challenges. Because of the special characteristic of energy conservation investment, there is a lack of awareness among non-core businesses and consumers. For the EMCos, who provide EPC services, the capacity to implement a simple project is needed, but more importantly, the capability to do an energy audit, design the project, and manage the construction are critical. Once implemented, the EMCo must oversee operations and manage the project as well as provide the measurement and verification of the energy savings. This is a much higher requirement for EMCos to minimize the risk and improve financial capacity. Obviously, the EMCos are facing more complicated and complex conditions.

Developing EMCos in China are facing financial, tax and payment terms, and other market barriers. The history of China EMCo development is the history of overcoming these barriers in order to further develop the industry. The EMCos in China creatively apply the EPC model to their businesses and have invented some new and comprehensive models, which have set up a very good example for other companies and also for the whole industry to study.

Main barriers for the development of ESCOs in China have been:

- The overall scale of the EMCo industry, which is relatively small compared with the great demand for the energy service, the existing industry is not big enough nor strong enough to meet the market requirements. There are not many large scale EMCos in the industry, most are SMEs.

- The capacity of the EMCos is limited. Faced with the increasing demand from energy consumers, the service ability of the EMCo still can't match the different and higher demand. Most of the EMCos lack financing methods, technical capacity and provision of solutions for customers. There is a long way to go before becoming good and strong professional EMCos.

- In large part, the "technical" EMCos lack the capability of technology innovation, staying "small but integrated." This prevents them from becoming broad-scale service providers. They are, instead, at the dangerous edge of being phased out of the market.

- Lack of professionals and talents: as the EMCo industry is a quite new industry, there are not enough professionals and talented people to meet the demand of this fast developing industry. This is another big bottleneck for the EMCo development.

- The market rules need to be set in order to prevent a competitive mess. The high investment returns and the great market potential attract more participants, which leads to faster development of the industry. But at the same time, some un-harmonized phenomena appear, which may influence the healthy and sustainable development of the whole energy conservation industry.

International Communications and Cooperation

As EMCA was started with an international co-operation project of the Chinese government with WB/GEF, communication with the wider world has never stopped. From the very beginning, in order to gain experience and learn lessons, and to improve the EMCo development in China, lots of activities have been carried out for information dissemination and co-operation. These include regular communications with ESCO associations in other countries, such as NAESCO in America, JAESCO in Japan, KAESCO in Korean as well as associations in European countries. It also involved irregular visits to other countries and internationally known ESCOs. The EMCA has been receiving more and more visitors from other countries, including Japan, Sweden, Norway, America, Germany, Britain, etc.

In 2007, the 2nd Asian ESCO Conference was held in Beijing, with about 300 representatives, mainly from Asian countries, but also from Europe and America, Australia, etc. This two-day event was a great gathering of ESCO experts, who shared the latest information on ESCO development in Asia. In order to further promote the EMCo development in China, EMCA will keep on strengthening international communication and co-operation for the sustainable development of the industry.

Summary: ESCOs in China

As with other industries, the energy service industry in China will meet new conditions, new problems, and new challenges during its development. If the new EMCos are to reach their full potential these problems must be solved and challenges met. The overall trend of healthy development of the industry will not change.

Chapter 7
North America

Conditions were ripe for the introduction of performance contracting in North America in the late 1970s. Energy prices were climbing. People were trying to implement ways to save energy, but expertise was often lacking. Needed funds to invest in energy efficiency were not always there, particularly in the public sector. Given these conditions, when Scallop Thermal brought the concept of guaranteed savings to North American shores, it was well received. Unlike the original French concept, which focused on the supply side of the meter, this new, guaranteed savings, approach focused on the demand side.

Canada and the US led the way in a resurgence of outsourcing energy management and the concept of energy performance contracting (EPC) spread quickly. This resurgence, however, met with two major obstacles. One was the lack of financing and the associated insurance some expected. The other was the legal profession, which was unfamiliar with performance contracts and many attorneys were reluctant to admit it. Another obstacle in the United States was the initial opposition of the federal government to the concept of performance contracting. All of these concerns were resolved over time, but it is helpful for those in other countries, in the early phases of ESCO industry development, to know that these obstacles did exist and at the time were rather formidable.

We are very fortunate to have Mr. Pierre Langlois, a pioneer in EPC in Canada, providing us with the particulars of how the concept evolved in Canada. After a brief history of its development and a summary of the ESCO industry's evolvement, he offers an excellent update on where the Canadian ESCO industry is today and what he foresees as its future. Mr. Langlois' perspective reaches well beyond the Canadian borders as he and his company, Econoler International, have offered ESCO development leadership in 35 countries around the world.

Mr. Donald Gilligan, president of the National Association of Energy Service Companies (NAESCO) with the assistance of NAESCO's Executive Director, Ms. Terry Singer, takes us down a similar path through

ESCO development in the United States. They offer a particularly detailed picture of where the industry is today.

Since Mexico's economy was not as strong as Canada's nor that of the US in the latter part of the 20th Century, it has only recently become active in any concerted ESCO development. Ms. Monica Perez offers us a very good review of an ESCO industry emerging in Mexico. Ms. Perez has taken an in-country leadership position in the International Energy Efficiency Financing Protocol (IEEFP) development and provides very valuable financial insights for other countries who wish to take advantage of IEEFP to adapt its framework to the unique needs of their respective business cultures.

CANADA

The rapid growth of energy performance contracting and energy service companies in Canada has been unique in that it was an early, government-inspired, solution, originating from the first oil price shock of the early 1970s and focused on inefficient use of energy and cost savings expectations.

The first ESCO in Canada and one of the first in the world was created in the province of Quebec by Hydro Quebec and a local engineering firm called ADS. Econoler Inc, was founded in 1981 and developed a new concept for the time, based on a unique shared savings approach with a "first out" option (an open book approach with contract termination upon complete payment of all the project costs, even if the contract period has not been completed) that was very attractive for the market at the time. The approach was all the more relevant at a time when interest rates in Canada for the implementation of such projects were in the vicinity of 20 percent, which significantly hampered investments, including those in the energy sector.

Between 1981 and 1989, in excess of 1,000 projects were implemented based on the first out concept in all kinds of commercial, institutional and industrial establishments in the province of Québec. The company invested over Can $135 million to implement these projects. Several of the clients financed them on their own, always, however, based on the energy performance contract approach. Annual recurring savings of Can $35 million were generated by the projects, which became recognized as very successful from both a technological and commercial point of view.

In 1983, following Econoler's initial success in the Province of Québec, the rest of Canada began to show interest in the new concept. Canertech, a subsidiary of Petro-Canada, government-owned at the time, wished to replicate the experience and make use of Econoler's know-how elsewhere in Canada. Econoler, therefore, signed a licensing agreement with Canertech that, in turn, transferred it to its newly created provincial subsidiaries located in Ontario, New Brunswick, Nova Scotia and Prince Edward Island. The subsidiaries were created between 1983 and 1987. Technical partners were local entities located in each of the provinces where the companies were launched. Starting in 1985, other companies, such as Rose Technology in Ontario, were created based on the same model, which created some competition in the Canadian market and launched what was to become a real ESCO industry.

Two major programs were launched in the 1990s to help the ESCO concept grow in the country:

1. The Federal Building Initiative (FBI), of the Canadian Office of Energy Efficiency, started officially in 1991, but the first implementation was done in 1993. The program assisted federal organizations in reducing energy and water consumption and greenhouse gas emissions in their facilities. The FBI, a voluntary program, addressed three common barriers to improve energy efficiency: (i) inadequate capital budgets for energy efficiency projects; (ii) need for reliable information on current energy technology and practices; and (iii) lack of required skills to manage retrofits.

    The FBI promoted energy services through publications, advice, including sample tender documents, information on energy efficiency-related environmental, health and safety issues, employee awareness products and comprehensive training programs.

    The program cost since 1991 is about Can $12 million. There are limited resources in the program unit; so every two years there is a bidding process to hire experts, who can provide support to federal organizations that want to enter the FBI program.

2. The Better Buildings Partnership (BBP) program began in June 1996 and focuses on cutting CO_2 emissions in Toronto through energy management firms' activities. BBP is a public-private partnership between Enbridge Gas Distribution Inc., the Toronto Atmospheric Fund, Toronto Hydro and Ontario Hydro Energy Inc. BBP is comprised of several programs including a residential energy awareness

program, an office building and commercial building program and a loan recourse fund.

In 1987, the Canadian Association of Energy Service Companies (CAESCO) was established with the support of Ontario Hydro, the federal and Ontario provincial governments. The association encouraged the orderly growth of the industry through accreditation, support and advice to both EPC contractors and customers. Membership in 1997 was over 50 and included, in addition to ESCOs, equipment suppliers, utilities, governments, lawyers and consultants. There are presently 13 accredited ESCOs. Several different level programs of the Canadian public administration to improve constructions' energy efficiency actually function owing to ESCOs. Unfortunately, CAESCO closed in 2001, for lack of support and interest by the different stakeholders.

ESCO Industry and Market

There are now 11 active ESCOs in Canada, even though an important number of smaller ESCOs operate in the market, mainly in specific provinces in the country. These ESCOs are listed as pre-qualified energy service companies through the FBI program and are listed on a qualified bidders' list available on the National Resources Canada (NRCAN) website. The program specifically allows ministries and governmental organizations at the federal level to contract with ESCOs through a predefined procedure.

The ESCOs' specific activities are poorly documented, as they are mainly operated by private companies, which are under private agreements except in the case of public sector projects. Since the break up of the CAESCO, it is not possible to have detailed insight on the ESCO market volume in Canada, as no one is keeping track.

The only real information available is based on the FBI program statistics compiled since the start of the project. To date, 85 projects in federal buildings are ongoing or have been completed, attracting Can $312 million in private sector investments and generating Can $44 million in annual energy cost savings. The total area of the 85 projects for which FBI projects have been implemented is estimated at 7.2 million m^2. These FBI projects have cut the greenhouse gas emissions by 285 kilotonnes annually.

Financing and Contracts

In Canada, the guaranteed savings model is used throughout the different provinces and it represents an important part of current contracts. As an example, the Quebec government, in the health and education sector, is favouring the use of EPC, but is financing all the project costs, requesting from the ESCO a turnkey approach and a savings guarantee.

To a lesser extent, Canadian ESCOs are also using shared savings concepts (in the FBI program among others) as well as the chauffage concept. Most ESCOs have access to pools of capital from private sector lenders or from their own organizations. ESCOs are able to structure the loan to be on, or off, balance sheet, depending on the tax situation and other considerations. Other financing options are also used, such as municipal or capital leases.

Companies, such as Capital Underwriters Corporation (a joint force of Canada's most experienced leasing and banking professionals established in 1992) offer a new approach on capital and fixed asset financing based on, or off, balance sheet treatment for both the energy service contractor and the customer.

More specifically, in the case of the FBI program at the federal government level, efforts were made to overcome budget constraints in the government sector by promoting private financing, either by an ESCO or lender institutions. The FBI program does not provide direct financing but creates a trust, which facilitates financing from a third party. The Government of Canada also offers incentives under other initiatives like the Canadian Industry Program for Energy Conservation (CIPEC).

Additionally, most utilities and some provincial energy efficiency agencies have a set of incentives targeting commercial, residential and industrial sectors, which bring partial financing for eligible projects.

Other special initiatives have been developed over time to help address the project financing issue. As an example, the Better Buildings Partnership Loan Recourse Fund provides loans, through Enbridge Gas Distribution Inc., to building owners for energy efficiency retrofit projects, by securitizing loans. This is an innovative program designed for the small/medium commercial buildings and the multi-residential sectors, non-profit or public sector organizations within the boundaries of the City of Toronto.

Types of Products

On a more qualitative basis we can say that since the beginning, the ESCO industry in Canada has been very concentrated in the public sector, and in the commercial building sector. This can be seen as a way of minimizing the risks involved in this business. Many of the public programs of the Canadian government, aiming at energy efficiency improvements in buildings, are managed by ESCOs.

In the early days the industry took a different turn from that in the U. S. In a typically Canadian way, the public sector got involved with the industry and this has continued in many ways so that a private-public "partnership" has developed.

Barriers

There are still some barriers existing in the market. One is the internal competition of public and private sector organizations, which charge fees to manage buildings but are not operating under an EPC concept. Those fees are based on a percentage of the operational expenses. Therefore, an energy efficiency (EE) project is a source of fee reduction for them. In the public sector, there are also some organizations that own an important number of facilities and can obtain capital budgets from the treasury council of Canada for building renovation; therefore, the ESCO concept is not attractive. Indeed, when an EE project is done by an ESCO, it reduces the opportunity to get the renovation capital and it minimizes the internal team and services that can be provided.

Even though the ESCO concept has been used in Canada for more than 25 years, there is still an important lack of awareness and knowledge in the market about the concept and how to use it. This is especially true in the private sector and even more in the industrial sector.

Enabling Factors

The Canadian ESCO industry has strong support from the government. The Canadian Government granted ESCOs immediate access to large public sector contracts. The credit rating of the government as a client made financing projects easy and the industry grew quickly. First out and open book-type (where all construction costs are bid and known by all parties) contracts prevail in the Canadian ESCO industry. This type of contract ensures greater trust between the ESCO and the client.

Through the BBP, the City of Toronto offers either a 2/3 interest-free loan or a guarantee equal to 50 percent of the investment for the implementation of ESCO procurements.

Other Canadian programs administrated by the Office of Energy Efficiency as well as utilities provide incentives and subsidies that strengthen the ESCO industry.

FBI Influence

There are not a lot of recent case studies available to demonstrate the actual ESCO market in Canada. Most of these case studies are made available through the FBI program. Below are four examples of such projects.

CASE STUDIES

Any relatively new industry can profit from examining the case studies of similar firms.

Table 7-1. Case Studies

Project Title	Outline of Projects
Banff National Park	**Client:** Parks Canada Agency
	ESCO: MCW Custom Energy Solutions Ltd.
	Measures: more efficient, higher-quality lighting, high-efficiency appliances, heating, ventilation and air-conditioning control upgrades, programmable thermostats and automated hot-water supply temperature settings, building envelope improvements, commissioning of mechanical and electrical equipment to maximize performance, skills-upgrading options for Parks Canada Agency employees.
	Contract: 10-year guarantee savings
	Project Costs: Can $500,000 to $900,000 by MCW
	Annual Savings (energy and water): 50,000 to Can $100,000 per year
	Emission reduction: 500 tons per year
Canadian Forces Base Valcartier	**Client:** Department of National Defence
	ESCO: Genivar
	Measures: Efficient T8 lighting with electronic ballasts, building management systems (30 buildings), upgrade the central heating system, new O2 remover for the boiler house and heat exchangers on the chimney, employee awareness campaign
	Contract: 10-year shared savings with a buy-back option starting from 1996
	Project Costs: Can $8.7 million using third party financing
	Annual Savings : Can $850,000
	Emission reduction: not available
Cityhome	**Client:** Cityhome Toronto
	ESCO: VESTAR
	Measures: Lighting retrofit, electric to natural gas dryer conversion, make-up air unit retrofit, domestic hot water heater conversion in 14 apartment buildings (1.65 million square feet)

(Continued)

Table 7-1. Case Studies (*Continued*)

Project Title	Outline of Projects
	Type of Contract: not available
	Estimated Costs: Can $1.22 million
	Annual Savings: 839,000 million kWh, Can $177,000, payback period of 6.9 years
	Emission Reduction: 2,013 tons CO_2 per year
Toronto Catholic District School Board Phase I	**Client:** Toronto Catholic District School Board
	ESCO: Ameresco Canada, Inc.
	Measures: Building automation system, lighting retrofit, water-efficient technologies and measures, heating, ventilation and air conditioning (HVAC) system upgrade, boiler retrofit in 45 school buildings totaling 2.6 million square feet.
	Type of Contract: not available
	Projects Costs: Can $5.8 million
	Annual Savings: 2.92 million kWh – Can $845,000 – payback period of 6.8 years
	Emission reduction: 2,786 tons CO_2 per year

Future Expectations

It is expected that the ESCO market will grow significantly in Canada in the next five years. Many provinces are promoting the EPC approach for the implementation of energy efficiency projects in the public sector. The FBI program has also remained quite active, and is expected to continue to promote its share of projects nationwide.

The private sector is timidly starting to look at the ESCO concept as a way to eliminate the different barriers for the implementation of EE projects. It is likely that this market will grow at a steady pace, as there is still an important market to be developed.

The rise of the energy costs everywhere in the provinces is certainly a huge incentive for the development of the EE market, and therefore of the ESCO market. The fact that the ESCO market is better known today than in its early days in the 1980s, will certainly play an important role in market growth in the near future.

THE UNITED STATES

This assessment of the ESCO market in the US has been digested from two reports and is presented by Mr. Donald Gilligan, president of the National Association of Energy Service Companies (NAESCO) with the support of Ms. Terry Singer, NAESCO Executive Director.

- *An Introduction to Performance Contracting*—principal author: Donald Gilligan, National Association of Energy Service Companies, as part of a subcontract to ICF International supported by the U.S. Environmental Protection Administration.

- *A Survey of the U.S. ESCO Industry: Market Growth and Development from 2000 to 2006*—principal authors: Nicole Hopper and Charles Goldman, Lawrence Berkeley National Laboratory; Donald Gilligan and Terry E. Singer, National Association of Energy Service Companies; and, Dave Birr, Synchronous Energy Solutions; funded by the U.S. Department of Energy

What is Energy Performance Contracting?
A typical energy performance contract (EPC) project in the US is delivered by an energy service company (ESCO) and consists of the following elements:

- **Turnkey Service**—The ESCO provides all of the services required to design and implement a comprehensive project at the customer facility, from the initial energy audit through long-term measurement and verification (M&V) of project savings.

- **Comprehensive Measures**—The ESCO tailors a comprehensive set of measures to fit the needs of a particular facility, its operation, and its owner, and can include energy efficiency, renewable technologies, distributed generation, water conservation and sustainable materials and operations.

- **Project Financing**—The ESCO arranges for long-term project financing, which is provided by a third-party.

- **Project Savings Guarantee**—The ESCO provides a guarantee that the savings produced by the project will be sufficient to cover the cost of the project financing for the life of the project.

Brief History of US EPC
The history of EPC in the US can be usefully divided into stages.

- **Initial stage (pre-1986)**—ESCOS grew out of the efforts made by Scallop Thermal, a Division of Royal Dutch Shell, as discussed in Chapter 1.

- **Oil price drop (1986)**—The embryonic ESCO industry almost died when the price of oil dropped in the late 1980s. At the time, all projects were using the shared savings model (a pre-determined split of the financial savings), which worked well when prices were stable or escalating. The 1986 price drop protracted the paybacks, often to periods longer than the contract. As a result, the US ESCOs shifted to guaranteeing the amount of energy saved and the guaranteed savings model emerged.

- **Public sector focus**—The initial focus of guaranteed savings was on the public sector market where the US tax-exempt financing made the model more attractive.

- **Expanding EPC (1987-1993)**—A boost to the industry came from utility programs, especially in response to demand side management (DSM) and Integrated Resource Plans (IRPs). ESCOs bid to provide the kW or kWh savings, delivered turnkey projects to large industrial and institutional customers and financed the projects themselves.

- **Success and Consolidation (1994-2002)**—Successful experience with EPC, documented in studies by the Lawrence Berkeley National Laboratory (LBNL) and the National Association of Energy Service Companies (NAESCO), encouraged the federal and state governments to promote the use of EPC in their own facilities as well as those of private sector building owners and operators. The implementation of the International Performance Measurement and Verification Protocol (IPMVP), which provided standard methods for documenting project savings, gave commercial lenders the confidence to begin financing EPC projects on a large scale.

- **Pause and then Fast Growth (2003-present)**—The collapse of Enron, the suspension of the federal ESPC program and uncertainty about the deregulation of the electric utility industry caused a US slowdown in the growth of EPC from 2002-2004. EPC is now growing at more than 20 percent per year, driven by increasing and volatile energy prices, federal and state energy savings mandates, the continuing lack of capital and maintenance budgets for institutional and federal facilities, and the growing awareness of the need for large-scale action to limit greenhouse gas emissions of which energy efficiency is the cornerstone.

US EPC Market Size and Characteristics

A recent study by LBNL and NAESCO has documented the current size and growth trends of the ESCO industry, as summarized in Figure 7-1 It shows a healthy industry growing at a steady pace.

In addition to size and growth estimates, the LBL/NAESCO report documented several other features of the ESC marketplace.

ESCO Ownership

As shown in Figure 7-2, the ESCO industry in the US has consolidated since 2000. Utility companies have largely abandoned the business in the US. About 80 percent of the total EPC business is performed by ESCO subsidiaries of large companies, primarily equipment manufacturers.

About three-quarters of the total EPC business is done by 10 nation-

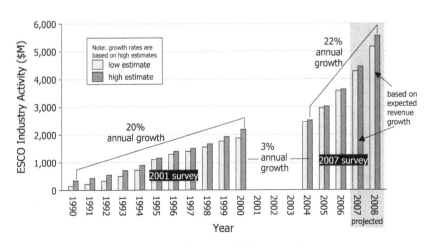

Figure 7-1. Industry Activity, 1990-2008

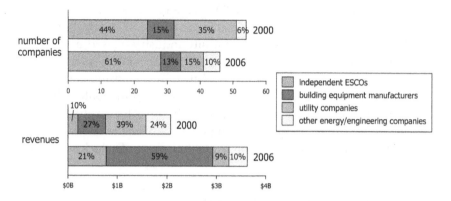

Figure 7-2. Trends in Industry Shares by Company Ownership

al ESCO companies, and another 20+ percent by regional ESCOs. Local ESCOs, who confine their activities to one or more local markets, do less than 5 percent of the national EPC business.

The MUSH (municipals, universities, schools and hospitals) market and the federal market account for about 80 percent of the total EPC projects in the US. Commercial building projects represent about 9 percent of the projects, industrial projects represent about 6 percent, and the balance consists of residential and public housing projects.

By dollar volume, ESCO projects are largely comprised of a mix of energy efficiency technologies (73 percent), renewable technologies (10 percent) and distributed generation or combined heat and power (6 per-

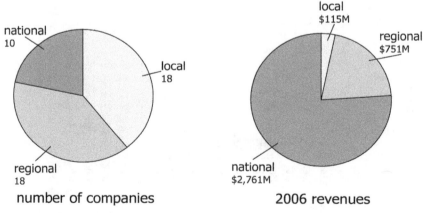

Figure 7-3. Industry Shares of Local, Regional and National ESCOs

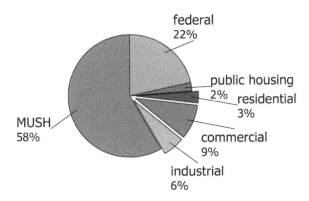

Figure 7-4. 2006 ESCO Industry Revenues by Market Segment

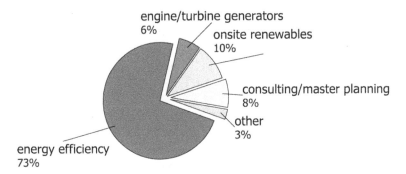

Figure 7-5. 2006 ESCO Industry Revenues by Technology/Project Type

cent). The balance of ESCO revenues is derived from consulting and planning services.

As depicted in Figure 7-6, about 70% of ESCO projects are performance-based, and another 25% of projects are implemented using design/build or engineering, procurement and construction (EPCS).

US Market Drivers

The US market has several major drivers, including:

- **Savings Mandates**—Federal and state governments are increasingly mandating aggressive energy savings goals for public facilities, but are not providing expanded capital budgets to pay for energy efficiency improvements. In this environment, EPC is viewed as a default method for implementing energy efficiency projects.

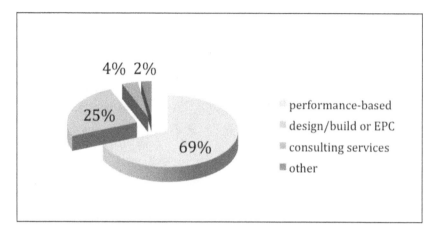

Figure 7-6. 2006 ESCO Industry Revenues by Contract Type

- **Facility Modernization**—MUSH (municipals, universities, schools and hospitals) market facilities, typically starved for capital and maintenance budgets, use EPC projects to obtain needed facility improvements without the need for directly allocated funding.

- **Green Buildings**—Facility owners who want to use renewables to "green" their buildings often implement EPC projects because "EE Pays for Green," that is, the savings produced from energy efficiency measures help to finance renewable measures.

- **Climate Change**—Energy efficiency is the first choice of policy planners trying to meet state mandates for greenhouse gas reductions. As with savings mandates, EPC projects enable facilities to meet greenhouse gas reduction mandates, which are not accompanied by capital budget increases to meet these mandates.

- **Utility and ISO/RTO Capacity Programs**—State regulators, faced with utility applications to build a new generation of power plants, are increasingly looking to large-scale energy efficiency programs as an alternative, and find EPC projects, which can be self-financed through energy savings, and deployed relatively quickly, attractive.

US EPC Financing

EPC projects in the US today are typically financed by third-party financial institutions, not by ESCOs, using a set of financing vehicles that are tailored to the requirements of an individual project.

EPC projects are mostly financed by large institutional lenders that offer very competitive rates and terms, and have made billions of dollars of financing available. The following list of financing vehicles offered by project financiers is adopted from *Performance Contracting: Expanding Horizons, Second Edition*, by Shirley J. Hansen.

- **General obligation bonds**—GOs are bonds secured by a pledge of a government agency's full faith, credit and taxing power. They are payable from ad valorem property taxes and require voter authorization. They are considered the most creditworthy by bond investors; and, therefore, are the least expensive form of financing.

- **Tax-exempt financing**—Available to entities that meet Section 103a, of the Internal Revenue Code. Organizations. Schools, which do not pay sales tax, are not necessarily tax-exempt.

- **Debt financing**—Direct loans from banks and credit companies may be straight debt financing. In such cases, the obligation to make debt payments is unconditional.

- **Lease-based financing**—From a legal perspective, lease-based financing differs from debt financing. As stated above, the obligation to make debt payments is unconditional. Lease payments, on the other hand, are conditional; i.e., they need only be made if the lessee has full use and possession of the asset being leased.

  Lease financing is the most common type of local government financing. "Municipal Lease" is a term often applied to leases even when a municipality is not involved. It is usually a simple lease and may be the nomenclature for tax-exempt entity and is frequently used as a legal definition.

- **Master lease**—This "umbrella" lease is a variant of the municipal lease with general terms and conditions. As the lessee makes individual purchases or begins individual projects, leases, or lease schedules, are funded and appended to the master lease agreement.

- **Certificates of participation (COP)**—This mechanism allows investors to purchase certificates, which offer evidence of their participation, and enables them to participate in the stream of lease payments being made by the lessee to the lessor. COPs have a much higher cost of issuance than municipal leases, but carry lower interest rates. They are well suited for pool financing, which can fund several projects. With low interest rates, they are usually fully committed with multi-year waiting lists.

- **Lease revenue bond**—Similar to a COP, except that instead of a corporation serving as a lessor, one government agency acts as lessor, often to a jurisdiction needing funding. The lease revenues are pledged as repayment of the bonds.

- **Power purchase agreements (PPAs)**, in which the customer buys the output; e.g., kWh or pounds of steam, of a distributed generation project, rather than owning the actual project.

US EPC Measurement and Verification (M&V)

The measurement and verification (M&V) of EPC project savings has evolved in stages which parallel the development of the EPC market outlined above.

- **Pre-1985**—Monitoring programs were initially used to track the progress of first-generation utility DSM programs, and tended to measure activities; e.g., number of audits delivered, rather than outcomes; e.g., kWh delivered. State energy officers, such as New Jersey and California, were also working to establish M&V protocols, but such protocols were not consistent from state to state.

- **1985-1993**—ESCOs and customers struggled to develop replicable M&V systems for unfamiliar technologies.

- **1994-present**—Successful project experience proved to customers that EPC projects involved little technological risk. Under US Department of Energy sponsorship, an M&V protocol emerged. First, the Building Energy Measurement Program, followed by the North American Measurement and Verification Program. In 1997, the International Performance Measurement and Verification Protocol (IPMVP) gave customers, ESCOs and institutional financiers a broadly accepted method for validating project savings and, with modifications, is accepted around the world today.

- **2003-present**—The emergence of various new EPC market drivers (see above) is pushing the development of an additional generation of M&V that will validate new streams of EPC project value, such as operations and maintenance (O&M) savings, greenhouse gas reduction and electricity system capacity credits.

US EPC Market Constraints

Several factors are holding back the growth of the EPC market in the US, including:

- **M&V limitations**—New systems are required to make the calculation of project energy savings more understandable to non-technical policy-makers depending on energy efficiency to meet public policy goals, such as energy savings and greenhouse gas reduction mandates. While M&V is viewed as a necessity, it is also seen as a cost burden to a project. Without major expenditures, the accuracy is only +/- 10-20 percent.

- **Shortage of skilled personnel**—ESCOs, utilities, state regulatory agencies and customers are struggling to find the skilled engineering and technical personnel required to implement large-scale energy efficiency and renewable energy programs, and to operate and maintain energy efficiency and renewable energy technologies.

- **Specific market barriers**—Each of the major EPC market segments suffers from it own constraints.
  - **The federal and MUSH markets** are hindered by landlord agency and financial control bureaucracies that too often resist large-scale program implementation in the face of executive and legislative mandates.
  - **The commercial real estate market** is hindered by the refusal of building owners to encumber their buildings with the debt required to finance comprehensive EPC projects, or by tenant problems.
  - **The industrial market** is hindered by the policy of most American manufacturing companies, which results in project payback requirements of typically less than two years, which preclude comprehensive EPC projects employing multiple technologies. Top management frequently fails to understand that financing

energy efficiency out of avoided utility costs is not the same as requiring new budget allocations. Fears related to the loss of intellectual property creates management reluctance as well.

Conclusions

The importance of the EPC market to utilities can be summarized with the major conclusions of the recent LBNL/NAESCO survey, referenced above, which stated, "ESCOs and EPC projects can be a crucial component of the rapidly expanding (in some states) or emerging (in other states) utility DSM programs."

Beyond the conclusions drawn from this survey, it can be said that ESCOs and EPC projects can be important contributors to the development of clean energy, sustainability and climate change mitigation strategies, particularly in urban areas.

US ESCO PROJECT CASE STUDIES

CASE STUDY #1: County of Fresno, California

Project Size: Several buildings totaling over 1 million square feet, plus a 1.25 MW combined heat and power facility
Project Cost: $12 million
Contract Term: 15 years—guaranteed energy savings program
Project Guaranteed Savings: $14 million total—$1.3 million per year
Rebates and Incentives: $1.5 million
Scope of Services: ESCO performed Investment Grade Audit (IGA) and implemented the project. The IGA included detailed surveys of all buildings, analysis and modeling of a comprehensive list of energy and water conservation measures and associated cost estimates. Implementation includes the design, installation and performance verification of the measures listed below.
— New 1.25 MW combined heat and power facility
— Conversion of electrical services from secondary to primary
— New power quality system
— High efficiency lighting modifications
— Upgrade chillers with variable frequency drives
— Upgrade and optimize the existing energy management system
— New air handlers
— New VFDs and high efficiency motors on pumps and fans

- Conversion of air distribution system and air handlers to VAV
- New high efficiency boilers
- New steam distribution system to optimize boiler performance
- New low flow plumbing systems
- New water treatment systems on cooling towers and laundry
- New irrigation controls

CASE STUDY #2: Industrial Process and Mechanical Upgrades
Marine Corps Logistics Base, Albany, Georgia
Situation: The Marine Corps Logistics Base (MCLB) in Albany, Georgia, provides worldwide, integrated logistics/supply chain and distribution management, depot-level maintenance, and strategic pre-positioning capability in support of military operating forces.

Executive Order 13123 mandates that federal facilities reduce their energy consumption 35% from a 1985 baseline amount by 2010. MCLB was investigating potential energy conservation measures (ECMs) that would help the facility achieve its reduction target.

Project Cost: $14,200,000
Total Energy/Operational Savings: $37,036,672
Equipment Installed/Services Provided:
- Implemented steam piping changes and installed remotely operable steam on/off regulating valves to isolate space heating loads and process steam loads
- Replaced existing steam turbine– operated feedwater pumps with new electric motor-driven units, and installed boiler blowdown controllers
- Installed three new "digital scroll" compressed-air dryers and water-cooled aftercooler and heat recovery equipment
- Installed 10,000-gal compressed-air receiver and demand controller
- Drilled 670 four-inch wells (approx. 250-300 feet deep) to install geothermal high-density polyethylene (HDPE) pipe loops with a 50-year life
- Installed 250 new geothermal heat pumps and electric water heaters and insulated attics to R-38
- Installed 494 natural gas–fired infrared heaters
- Installed a web-based direct digital control system to monitor and control building HVAC equipment in 39 buildings

- Installed one 60-ton and one 215-ton VAV system with new chillers and controls

Solution: Under the DOE's Southeast Regional ESPC program, MCLB awarded a contract to an ESCO to implement ECMs as part of a bundled Delivery Order project. ESCO performed a comprehensive energy analysis that investigated potential ECM options in 12 technology categories within approximately 105 facilities. The audit considered intensity of energy use, facility square footage, and recommendations made by MCLB personnel.

Eight ECMs were deemed viable and selected for implementation:
- Steam distribution and maintenance
- Compressed air and heat recovery
- Geothermal heat pumps
- Infrared heat
- Lighting upgrades
- HVAC equipment replacement
- HVAC renovation
- Web-based DDC

Project Benefits: In addition to reducing energy consumption by approximately 167,125 MMBtu, the improvements installed at MCLB are designed to provide several key economic, infrastructure, and comfort benefits:
- Energy savings and maintenance are locked in a long-term agreement. This ensures performance, reliability, and energy savings even if there are cutbacks in base maintenance budgets or personnel reductions.
- Significant improvements in steam and compressed air production and distribution and waste reduction has allowed depot maintenance activity to double production and personnel in response to Operation Iraqi Freedom without increasing the number of boilers or air compressors.
- A 100,000-sq-ft headquarters building was renovated, increasing the number of HVAC zones from 9 to 72, which has improved humidity and temperature control and occupant comfort.
- Under the ongoing services program, ESCO will routinely inspect installed equipment and perform preventative maintenance, ensuring reliability and long-lasting performance. An annual steam sys-

tem audit and repair program will keep the distribution system in top condition thus reducing energy loss.
- A new 10,000 gallon compressed air receiver tank with demand controller and state-of-the-art air drying equipment reduces false demand, excessive compressor cycling, and energy per unit of air produced, while improving reliability.

MEXICO

In Mexico, by the end of the 90s, several engineering and consulting firms realized that there was an increasing need in the market for an integrated services approach carrying both the technical and financial support required to implement comprehensive energy efficiency and renewable energy savings based projects (ESPs). Ms. Monica Perez offers a broad overview of the conditions that spawned ESCO development in Mexico, a review of the current status and an excellent examination of the financing situation that has plagued ESCOs in many countries. As the Mexican lead in the development of the International Energy Efficiency Financing Protocol (IEEFP), she shares the ways in which IEEFP might assist ESCOs in other countries.

Mexican ESCOs offer identification, development, financing, installation and operation and maintenance services at end user facilities on a performance basis. The energy end user is offered no-cost financing, which will be paid out of the savings generated through the implementation of the measures in its facilities. The owners also receive additional benefits, such as reduced operating and energy costs, new equipment, increased competitiveness and the positive impact of becoming an environmentally concerned business.

In order to offer this type of service, Mexican ESCOs have accessed resources from private funds or invested their own, family or friends' money in the business. Most Mexican ESCOs offer a shared savings structure; however, a couple of well established ESCOs with good track records are able to get energy end users' commitments to pursue the guaranteed savings structure. For nascent markets, such as that in Mexico, the shared savings structure has been one of the features that enable ESCOs to engage end users into entering such innovative financing mechanisms. Having a model in which the end user does not need to invest in project implementation, nor to access debt, and get shared savings from the beginning, has often been perceived as being too good to be true.

Until 2005, some 10 ESCOs were actively being supported by the U.S. National Renewable Energy Laboratory (NREL) and Mexico's National Commission for Energy Conservation (CONAE) Collaboration Program, through catalyzing activities and promotional events with industry, hotel organizations and targeted associations. Activities included site visits at end users' facilities for project identification, promotion of partnerships with international companies, etc. Several consulting and engineering firms showed strong interest in becoming ESCOs, but the transaction costs of acquiring the technical and financial know how, along with accessing financing resources, made it very difficult for them to turn into ESCOs.

Although there is current government support to develop tools to facilitate ESCO market deployment, there are several market barriers that hinder development. These include lack of awareness, high transaction costs, lack of tax incentives, current procurement rules that are disincentives to ESP development, unknown size of the market, and problems accessing financing, etc.

Similar to what has happened in other countries, even for established ESCOs, accessing financing has proven to be difficult enough to have an effect on the growth rate of the industry. With the financing structure of shared savings, the ESCO carries the financial burden on its balance sheet and, therefore, after financing some projects there is no additional debt capacity left for them to pursue additional projects.

The size of projects in Mexico range from USD 100,000 to USD 5 million, which presents a problem, since these projects are too big for micro finance programs and too small for typical transaction costs.

Current ESCOs Industry and Market

In addition to true ESCOs, the Mexican industry is a mixture of equipment suppliers and engineering companies.

There are a few well established ESCOs, which truly offer comprehensive ESPs. These need to take into consideration how measures interact within ESPs. For example for cooling dominated weather, where the air conditioning load is considerable, the ESCO could offer use of special glazing units, which could have a great impact on reducing cooling loads. At the same time, it could also reduce daylight entering the building; hence, the lighting would have to be increased. For projects in the industry sector, this interaction is much more complex and meaningful since it affects processes. Not too many ESCOs have enough technical capacity and technology partners to include comprehensive measures with a positive interaction.

Most of the ESCOs focus on a specific technology where they have either mastered their skills, have a joint venture with the equipment manufacturer, or are themselves the manufacturers. This approach, however, could present a problem, as the equipment manufacturers may be trying to sell as much equipment as possible instead of providing the elements required for an integrated approach.

Financing

An assessment of current ESCO conditions in Mexico related to financing provided the following observations.

- The most typical financing sources for performance-based projects in Mexico are ESCOs, the client's own money, participation of private trust funds, development banks and commercial banks.

- ESCOs offer energy end users the implementation of ESPs based on a performance contracting or outsourcing basis. ESCOs participate, either with their own money, private equity funds, and/or leasing structures. The shared savings approach is the most common practice among Mexican ESCOs, however, projects based on guaranteed savings are also sought by a few well established ESCOs.

- At a 2004 ACEEE summer study, Cohen, Renne, and Perez Ortiz observed, " ESPs require up-front investment with economic return depending upon the performance of the project. Because ESCOs are often unable to demonstrate positive profitability of their company in earnings reports in the first period of the contract, banks consider these operations risky and request collateral. No debt capacity is likely to be available in order to access more funds after financing a few projects."

- To access financing from equity funds and reduce transaction costs some ESCOs have established special purpose entities. Only few ESCOs with very good projects, which can cover a 25 percent after-tax ROI, have been able to access these resources. This is not now attainable by the critical mass of ESCOs or engineering companies willing to become ESCOs.

- *Fondelec Latin-American Clean Energy Services Fund (FLACES) and the Clean Tech Fund* (owned by Econergy International Corporation which recently acquired the Mexican ESCO Empresas ESM) are private equity funds. FLACES has backed up Mexican ESCOs through SPEs especially in the hotel sector. This equity fund is closed now. They have, however, attained very good knowledge of the Mexican market and could play an important role in the future. The Clean Tech Fund has used the majority of its resources in Brazil.

- *The Trust Fund for Electric Energy Savings (FIDE)* is a non-profit private organization, which receives funding from Comisión Federal de Electricidad (one of the electricity utilities) for operation. FIDE promotes the rational use of electric energy for the private sector. FIDE provides support through energy audits, training, financing, technical assistance, replacement of high efficiency equipment, etc.

  One of the special features that have brought a positive repayment rate is that FIDE receives payment of loans through the electricity bills. FIDE receives resources from Nacional Financiera and has been very successful in implementing electric energy projects but leaves the thermal component of energy savings projects out of the spectrum for loans.

- *Trust Fund for the Thermal Insulation in Houses (FIPATERM)*. The objective of this project is the installation of energy efficiency technologies for the residential sector, mainly reducing air conditioning loads. This project targets some states in the north. The first stage of this program is targeted at the replacement of roof insulation material and the installation of efficient lighting systems. For the second phase the program includes additional technologies such as refrigerators and motors. The program, renamed Systematic Integral Savings (ASI), covers additional states.

- Financing exists but is not always accessible. Nacional Financiera (NAFIN) has an agreement with the Export Import Bank of the USA, to offer the Master Guarantee Agreement (MGA). This program was perceived as very attractive by some ESCOs since NAFIN acts as a first floor/commercial bank by providing financing directly and eliminating the intermediary. ExIm Bank provides the guarantee. This program was not designed specifically for performance-based projects although it could have been very effective for energy efficiency financing. However, the selection criteria used by Nafin and ExIm bank proved to be very restrictive and even the most well established ESCOs have not been able to access these funds.

- *NAFIN* also has the SME Guarantee program where it guarantees commercial banks a percentage of the loans granted to SMEs for working capital or fixed asset investments, which can be used to complement collateral that creditors must provide to banks. The bank's access to NAFIN provides an automatic guarantee up to an amount "awarded" to a bank through auction process. NAFIN evaluates the bank's "loan granting model" to be included in the list of intermediaries who could operate the guarantee program. The bank's model is a "scoring" (Parametric) type of model.

- *National Bank of Public Works and Services (BANOBRAS)* is a development bank, which promotes and financially supports energy efficiency projects. They focus mainly on public lighting, water pumping, lighting in public facilities and renewable energy use for electric energy generation for self-supply. BANOBRAS signed an agreement with CFE, CONAE, FIDE in July 2001 in order to develop energy savings projects in states and municipalities. Credit is available for private and public sectors.

- *North American Development Bank (NADBank)*. This is a bilateral development bank, which offers financing for the border region between Mexico and the US—300 km from the border line into Mexico and 100 km from the borderline into the US. This bank has been the implementer in the ESMAP Innovative Financing for Energy Efficiency in Mexico. This bank offers direct loans, technical assistance and guarantees for project development in public and private sectors.

Based on the experience of interviewing 6 local banks in order to develop the Energy Efficiency Financing Assessment Report for the IEEFP-Mexico, it can be noted that,
— The participation of commercial banks in ESP financing has been very marginal. They have managed to participate in project finance for large renewable projects with a considerable size (USD 50 to USD 100 million). The floor and limits for these projects are not consistent with the ones required for ESPs, which are around

USD 500 K to USD 2 million. This restriction leaves out the majority of energy efficiency projects typically offered by energy service providers and project developers.
— Nonetheless, some banks have participated in projects with lighting systems, air conditioning units, thermal glazing units, peak shaving, boilers and small hydro projects. In these cases, banks have tailored the financial structures according to the needs of the energy end user and the project. The learning curve has been so steep and associated costs have been so extensive that the experience has been negative. As a consequence, some banks have decided not to continue financing ESPs until the ESCO market is more mature and solid.
— There are no specific lending practices for ESPs. Banks use the same traditional selection criteria which is very restrictive.
— Commercial banks do not have technical departments or experts who can readily assess the real risks and benefits of ESPs, therefore, they tend to allocate a greater risk to these projects in order to grant a loan.

Although carbon market brokers are interested in energy efficiency projects which reduce greenhouse gas emissions and qualify for Certificates of Emissions Reductions (CERs) for the Clean Development Mechanism (CDM) of the Kyoto Protocol, or Verified Emissions Reductions (VERs) for the Voluntary Markets (Europe, Japan, USA, etc.), these projects are often too small to justify the high transaction costs of crediting the emissions. Bundling several ESPs either by energy end user or by ESCO could help reduce transaction costs and still provide a profitability margin.

Some of the recommendations from the Energy Efficiency Financing Assessment Report point out that the financing needs for Mexico should be provided by local financial institutions (LFIs) on a project basis as noted below:
— "No collateral requirement beyond EE cash flow and equipment from developer or creditworthy hosts;
— Minimum 7-yr. term + construction period;
— Construction financing provided;
— Repayment in local currency;
— "All in" costs must relate to market interest rates; and.
— Must be able to clearly measure & verify savings."

Types of ESCO Projects

Most of the Mexican ESCOs have focused on specific sectors, wherein their expertise has a significant effect on the energy consumption and/or water consumption reduction. For example, one firm has successfully developed projects for the commercial sector, especially hotels and hospitals. In these cases, technologies focused on heat recovery systems, sea water for cooling systems, lighting, peak generation, etc., are employed.

Empresas ESM and Energy Savings de México are the two ESCOs, which have mainly targeted the industrial sector, although they have not been involved in projects that considerably affect the industrial processes. Diram has also focused in the industrial sector and has developed performance contracting in energy related projects. Even though they are not considered energy efficiency projects, they have an impact on the reduction of the energy bills and are projects very easy to verify through proven technologies as well as the energy bill.

In the case of Ecotherm, which develops projects and sells equipment related to water heating systems, heat recovery, heat economizers, boilers, etc., the firm has mainly targeted hotels, jails, and recreation facilities, such as sport clubs.

Guascor has been developing projects more on the side of generation. Project sizes are in a different range than those of the rest of the companies listed here, which has influenced its access to resources and the focus of its markets.

Secner has focused on the retail stores market with air conditioning, lighting and refrigeration projects. Novaenergia has targeted the hotel sector with power factor correction and demand control for process industries (heat pumps). Even though these two companies do not have an extensive track record as some described above, they are negotiating some projects and strengthening their offerings and will soon have successful stories to back up their efforts.

Barriers

A number of barriers to ESCO development in Mexico have been identified. Some of the more significant barriers are described below.

Accessing Financing

Accessing project financing by project developers and ESCOs has been particularly difficult. The perception of high risk relating to ESCO projects and the conservative lending practices previously described, have left ESCOs with restricted access to loans. ESCO projects require up-front investment and the economic return comes periodically depending on the performance of the project. In Mexico, a contract term can last from a year to 10 years. When reviewers examine the creditworthiness of the ESCO, they look primarily at their balance sheet, which does not reflect an immediate return of the investment. This makes ESCOs the least desirable candidates for credit allocation. Since the Mexican ESCO may not be able to demonstrate a positive profitability in their earning reports in the first period of the contract, banks consider these operations risky. As a consequence, they request collateral to cover the perceived risks, which in most cases, the ESCOs have been unable to provide.

Some companies have been able to secure financing with Trust Funds and special credit lines. After financing a few projects, however, they have no debt capacity available to access more funds. Due to this lack of equity, ESCOs have not been able to continue engaging in more projects, as they cannot qualify for further credit. Several ESCOs have been forced to invest private resources in order to continue project commercialization.

One of the most significant barriers to the widespread implementation of clean and "proven" energy efficient technologies around the world is a lack of commercially viable financing. The *problem is NOT a lack of available funds, but rather an inability to access available funds* at local financing institutions (LFIs). This problem is *caused by a disconnect in current lending practices* of LFIs and needs of energy efficiency & savings-based renewable projects. LFIs typically provide "asset-based" lending, which is limited to 70 to 80 percent of the value of assets being financed (or collateral being provided), and they do not acknowledge (or believe) that meaningful cash flow will be generated from ESPs, nor that such cash flow will equate to increased credit capacity for the end-use energy consumer. This is caused in large part by the fact that LFIs are not familiar with unique intricacies of ESPs, and they do not have the internal capacity to properly evaluate the risks/benefits of ESPs. Furthermore, due to the relatively small dollar size of ESPs, LFIs are unwilling to invest in building internal capacity to properly evaluate ESPs.

High Transaction Costs

Promoting and implementing ESPs have considerably high transaction costs, which result in increased risk and high interest rates. In order to sell an ESCO project, ESCOs need to have a solid, integrated offer and know how to approach the client. Some Mexican ESCOs or engineering companies willing to become ESCOs have a very technical versus a business approach.

Since the market is still very new, there are no standardized documents, such as performance contract or energy services agreements (ESAs) or traditional audit reports, Investment Grade Audit (IGA) reports, or even a series of elements that these documents might integrate in order to have a successful project. There is a clear need for legal advice to prepare these documents, which can become very costly.

Measurement and verification of savings, which is critical for ESCOs implementation is not a standardized practice in Mexico. Although there are international tools, such as the International Performance, Measurement and Verification Protocol, Mexican ESCOs have not regularly used it as the IPMVP appears to them to be a costly process and that Mexican conditions are not very applicable. They have, therefore, been establishing their own monitoring and verification methodologies, which have resulted in increased transaction costs.

Another factor that affects the transaction cost is the cycle to evaluate, negotiate and sell a project. This period could go from one to three years and sometime in the middle, the potential customers might get tired, or the financier lose interest in pursuing the project.

As indicated before, USD 100,000 to USD 5 million projects are considered too small in relation to the transaction costs. Considering all the risk involved, the high interest rates and all the factors mentioned above, some projects are not attractive and profitable enough to be economically viable.

Lack of Awareness

Many energy end users do not understand how these projects work. They often think that they are simply too good to be true. Sometimes energy end users are reluctant to share information crucial to their businesses. Or, they are reluctant to allow a third party to participate within their facilities and possibly negatively impact their industrial processes.

Whenever negotiating a project, sometimes end users realize the profitability of the project and, ironically, prefer not to enter into a contract; thus, continuing to waste energy and money instead of sharing the profit with an expert company.

Finding the right partner within the facility to move a project forward can be a problem. Some representatives of the technical departments feel that the ESCO is proposing business opportunities, which he/she should have identified in the past; and instead of embracing the project and becoming a promoter, they would rather block it. It must be noted that having the financing departments or decision makers' buy-in is crucial from the very start of the project. Legal and technical personnel need to be involved in parallel.

Unknown Size of the Market

There is still not yet an estimation of the size of the Mexican ESCO market and, not having a specific market potential has made LFIs somehow skeptical about the feasibility of ESCOs. Having successful cases, however, can be the way to show all the potential target actors the effect of ESCOs. There are several international experiences that have proved the case regarding the feasibility of the financing structures; nonetheless, specifics relating to the Mexican market are still pending. Mexican ESCOs have provided some information about successful cases, but as attaining their "know how" has been the result of a long, costly, and cumbersome process; they are often unwilling to share much information.

**Regulatory Constraints for ESCO Project
Development in Federally Owned Facilities**

Mexican regulations have inhibited energy efficiency project development in the public sector due to the regulation that prohibits entities from entering into long-term financial obligations without the authorization of the Treasury (Secretaría de Hacienda y Crédito Público). Although it is possible to get authorization, the process is complicated and time consuming. However, some recent work by the Investment Unit of the Underministry of Expenditures of the Treasury (which has imported from the UK the Private-Public Partnership or PPP) could provide the framework, which ESCOs need to enter the public sector. Even though there are some differences between types of projects that would fit into to PPP, this could be the key to help introduce the ESCO concept to the federally owned facilities and implement some projects. Once the parties involved are more comfortable with this financial alternative, a suitable framework for ESCO project development could evolve.

Public procurement rules are a disincentive for ESCOs. If an ESCO develops a project for a public facility and the energy consumption bill is reduced, the following year the expenditure budget would have to be reduced as well and so on for the following year. According to the current regulations, it would not be possible to apply extra funds resulting from the implementation of the measures, to another use under the budgetary structure. Such rules remove the owner's incentive to participate and removes the funds needed to pay the ESCO.

Non-existent Tax Exemption Schemes

Although there are some tax deductions relating to investment on some equipment that utilizes renewable energy sources and/or conversion of fuels, there are no current tax incentives or rebate programs that regulate energy efficiency. Therefore, there is a great opportunity to develop joint agency programs to address this barrier.

Enabling Factors

There are several initiatives supported by governmental institutions and international organizations oriented to foster ESCO market development in Mexico. These include capacity building of local financing institutions and other stakeholders, as well as providing additional elements as a way to help overcome barriers and reduce transaction costs.

The International Energy Efficiency Finance Protocol for Mexico (IEEFP-Mexico) is a series of tools and best practices that will be used as a blue print to train LFIs on how to evaluate and assess the real risks and benefits of ESPs. IEEFP is an international effort, which is part of the work of the Efficiency Valuation Organization (EVO), which also issues the IPMVP. The IEEFP Committee is chaired by Thomas K. Dreessen, an early pioneer in ESCO development. The project was initially funded by the Asia Pacific Economic Cooperation (APEC) and is currently supported by the Global Opportunities Fund of the United Kingdom.

The main goal of the IEEFP is for LFIs to have a better understanding of how the cash flow of ESCOs can become the main repayment source for ESCO loans. In the long run, this program is expected to result in a considerable increase in participation of LFIs in ESCO financing due to a reduced risk perception. Also, several other stakeholders will become more involved and interested in this market once there are standardized practices with decreased risk and transaction costs. The market will become sustainable and would not be in need of support programs.

IEEFP-Mexico has been developed with full participation and support of the Ministry of Energy (SENER) and NAFIN along with several other market stakeholders, which has integrated the Economy Working Team (EWT) including government agencies, banks, project developers, engineering firms, private funds, etc. This program has been launched nationally and has included two training courses recently conducted in Mexico City and Monterrey.

Feedback and interaction of LFIs has been sought at all moments in the program development in order to ensure that the program responds to the needs of the market. Trainees of the pilot training acknowledged that the course has "positively changed their perception about financing ESPs, and they admitted there is currently room to finance ESPs in their respective financial institutions." They all said that training should be launched nationally in order to change LFIs' vision on energy efficiency and to work in parallel with certain guarantee mechanisms being offered by NAFIN in order to help catalyze the financing of ESPs.

There is a critical need to have measurement and verification protocols involved to assure savings are estimated and achieved properly. Many Mexican ESCOs have developed their own measurement and verification methodologies, which might be specific to their technologies and projects but are not universal and easy to replicate for other projects. Having a standardized method, which is already internationally accepted, could be more effective and reduce uncertainties by providing more transparency to the project.

In 2006 EPS Capital developed a study for NAFIN and the US Trade Development Agency. As a result of this study, the firm made the following three recommendations to NAFIN in order to reduce the barriers to accessing financing of ESPs:

- Provide a performance risk guarantee to cover the savings shortfall to LFIs.
- Provide a partial credit guarantee to cover the credit risk of the energy end user.
- Use the IEEFP and Special Purpose Entities to bundle financing of ESPs.

The implementation of such an initiative would enable LFIs to build internal capacity to finance ESCOs in such a way that they feel comfortable enough with the risk; thus, reducing transaction costs, encouraging project bundling and helping to create a sustainable market to finance ESCOs.

Energy Sector Management Assistance Program (ESMAP) Innovative Financing for Energy Efficiency from the World Bank has program goals similar to the IEEFP, since it seeks to lower transaction costs through standard transaction documents. Stated ESMAP goals include: "The objectives of the project are to: (a) identify interested lenders and borrowers and potential projects to bring into the pool; (b) create a special purpose entity or intermediary to manage the process and represent a pool of project to banks and other sources of capital; (c) lower transaction costs by developing standard transaction documents and operating procedures for the Special Purpose Entity (SPE); (d) mitigate risk by spreading the lending over a portfolio of qualified projects and connecting the portfolio to available credit enhancements; and (e) improve the comfort of both borrowers and lenders by creating a process that is standard, transparent and more predictable." The North American Development Bank was the implementer of this project in Mexico.

The ESCO Market in Three to Five Years

The Mexican ESCO market is in a period in which government agencies, LFIs, project developers, as well as private financing funds are working together in order to identify the ways to address the barriers and facilitate the financing environment to bring about a sustainable market.

Elements, such as tools to evaluate and assess real risks and benefits; e.g., the IEEFP, will help LFIs get comfortable with the ESCO and energy performance contracting concept. This will assist LFIs to start considering cash flow as the main repayment source for energy savings based projects. Along with an increased low risk perception, which is expected as a result of NAFIN's guarantee mechanisms, it is hoped that LFIs will see ESCOs as replicable sustainable businesses. In five years, it is hoped that when an energy end user asks a bank lending officer for information about ESCO financing for its medium to small enterprise, the lending officer will know what that means.

The possibility of establishing an ESCO association in Mexico has been proposed in the hopes that, by sharing information and experiences, such an organization would enable:
- The needed changes in areas such as legislation;
- Procedures for participating in federal activities;
- Creation of new financial products by the LFIs;
- Establishment of tax incentives and rebate programs; and
- Application of standard protocols oriented to increase confidence and transparency to the ESCO's process.

Although potential for the combination of energy efficiency and renewable energy has not been assessed yet in Mexico, it is known, given the considerable inefficiencies in the country's industry, there are great opportunities for financing mechanisms that could have a meaningful impact on energy consumption reduction.

CASE STUDIES

The following pages were extracted from the *Energy Efficiency Financing Assessment Report* developed under the International Energy Efficiency Financing Protocol for Mexico (IEEFP-Mexico) and are descriptions of some of the ESCOs that were promoted by the NREL-CONAE collaboration and which provided input and support to the IEEFP development. Only the most relevant ones are included in this section. There might be additional companies currently offering ESCO services that have not approached the national entities promoting the market.

DIRAM

Services Provided:
- Areas of expertise: ESCO that implements paid-from-savings EE projects within core EPC business. Focus on electro-technologies. Primarily implements power factor reduction.
- Market focus (sector): industrial.
- All expertise in-house or subcontracted.

Size:
- Sales/Yr.: $10 + million USD (100 million pesos).
- Staff #: more than 20 people.
- Projects completed: 250 in total.
- Average Project Size, $200,000 USD ($2 million pesos).

Ownership:
- Independent company, privately owned.

Description of Financing Offered:
- Self-funds as many projects as possible out of working capital.
- Also finances projects with local private investors funded with equity that requires 25% after tax ROI and only reimburses Diram its costs and no profit until equity is repaid in 2 to 4 years.

ECOTHERM

Services Provided:
- Areas of expertise: Sell and install their line of products: economizers, eco-heaters, high-speed exchangers, solar heating technology, energy recovery, boilers, cogeneration equipment, and heat pumps. They focus on thermal energy projects and are also looking at thermal energy in greenhouses. They combine two schemes, can sell and install their equipment or develop projects based on performance contracting.
- Market focus (sector): hotels, sport clubs, jails, universities, etc.
- All expertise in-house or subcontracted. They have all the expertise in-house, considering that it is an Austrian company with presence in several countries.

Size:
- Sales/Yr.: In Mexico, sales are not too good compared to other economies where they have presence. Approximately 1-1.2 million USD per year (10—12 million pesos per year).
- Staff #: Internationally, including Austria and Mexico, 80 out of which 11 is Mexico's staff. They are present in 30 countries.
- Projects completed: 40 out of which 80% have been traditional sale of equipment and 20% performance contracting (they started operations in Mexico in 1993).
- Estimated Growth rate with favorable financing: Their annual growth rate goal is 20% although sometimes they are not able to meet it.

Ownership:
- International Austrian Company.

Description of Financing Offered:
- Traditional sale or performance contracting with access to credit lines of up to 10 million USD.
- Structure of the ESCO project. Once they have historical consumption, they make projections with financial runs and determine what the finance terms might be. There are two schemes; they either pay part of the total savings or a percentage. Most of the projects have a shared savings structure of 80% (ESCO)—20% (client). Once the historical data are established they are integrated in the contract. Since they finish installing in the middle of the month they usually give

the client a grace period and start charging from the start of month. They have a pure leasing structure due to Mexico's regulations and in the end of the contract they have a buy-out option.
- Sources: This company has agreements with Austrian banks that give them open credit lines. There are no agreements between Mexican and Austrian banks for these projects, so the company has to do it directly. In other countries there are agreements between banks. Since there is a lack of financing structure in Mexico, they allocate the resources directly without any participation of the Mexican banks. In other countries such as Germany, Belgium and Holland, they have very expedited agreements among banks.

ENERGY SAVINGS DE MEXICO

Services Provided:
- Areas of expertise: Electrical energy efficiency project, thermal, peak generation, air conditioning, lighting, demand side management, ice storage, energy quality, depressurizers.
- Market focus (sector): industrial, institutional, and commercial.
- No subcontracting, all work is done with in-house expertise.

Size:
- Sales per year: 2 million USD ($20 million pesos).
- Staff #: 35 employees.
- Project investment: from $5,000 pesos to $5 million USD ($500 USD up to 500,000 USD).

Ownership:
- Local independent company, privately owned. Has an alliance with a Spanish company for the depressurizers. They are looking for additional investors.

Description of Financing Offered:
- Performance contracting basically. Shared savings 80-20. Also traditional consulting and sales of equipment.
- Contract reflects a leasing structure.
- Own sources. Have not been able to access international funds.
- They use FIDE when the client wants to access their resources. Have used these resources for AA units, depressurizers, etc.

SECNER

Services Provided:
- Areas of expertise: Integral energy savings projects in lighting, air conditioning and refrigeration; savings are guaranteed by a performance contracting policy. They offer monitoring services for management. They have been operating 2 years with this scheme.
- Market focus (sector): heavy industry, water organizations, self-service stores, airports.
- All the expertise is in-house. They have an exclusivity alliance with the company III which has extensive experience in the energy sector and has worked for over 30 years. III is in charge of designing all the engineering.

Size:
- Sales/Yr.: 2005 was the first year and focused on promotion. 2006, 1 million USD.
- Staff #: 10 people.
- Of project (investment): From 200,000 USD to 10 million USD (from $2 million to $100 million pesos).
- Estimated Growth rate with favorable financing: Approximately 10 times.

Ownership:
- Local Public Limited Company. Has a joint venture with III for all projects.

Description of Financing Offered:
- They are currently looking for financing.
- They offer shared savings with profit over the cash flow.
- Currently they are evaluating the chances with private funds.
- Projects have been funded with their own resources.

Chapter 8
South America

As a continent rich in talented, vivacious people as well as great natural resources, South America seldom receives the regard it deserves. It could well be the sleeping giant of ESCO opportunity.

This chapter presents an intriguing cross section of ESCO history, development, and current conditions from three countries. We are fortunate that the government of Chile asked Karin Gauer to provide us with a description of its ESCO situation. Marcelo Gonzalez presents us with a brief synopsis of Uruguay's ESCO development and current activities. Offering an inviting glimpse into Brazil's evolving ESCO market, Alan Poole sets the stage by captioning his section, "Challenges of a Favorable Moment." Brief biographies of these recognized experts appear in Appendix A.

CHILE

Until recently, high availability of hydroelectric power and the supply of cheap natural gas from Argentina kept electricity and gas prices low, discouraging investment to improve energy efficiency. Some isolated efforts were made by the private sector as a result of the modernization of equipment, especially electrical and combustion engines. These projects were primarily in high energy consumption industrial sectors, such as the copper mining, metallurgical, cement and forestry industries. However, these efforts were undertaken as part of internal company policies, and did not involve external support, such as an ESCO.

Nevertheless, within this context, initiatives were launched to provide energy efficiency services. In this sense, it is worth highlighting the creation of Compañía Nacional de Energía (National Energy Company, CONADE) in the 1980s as a result of the forward-looking vision of an entrepreneur in this sector. At that time, the entrepreneur was selling heating and steam to numerous shopping malls that were being built in all major cities.

Another important attempt to create ESCOs in Chile occurred in the mid 1990s with the arrival of a French group, Vivendi (which today is Veolia Environnement). Attracted in 1995 by an open market and Chile's sustainable economic growth, this group created ESENER S.A. (a subsidiary of Dalkia, part of the Veolia Environnement group) as an ESCO, offering services related to outsourcing the operation and maintenance of energy facilities.

The purchase of CONADE by Dalkia in 2004 meant a great leap forward and led to the creation of the first line of business in Chile of different types of energy efficiency service contracts for industry and commerce. Dalkia, which combined ESENER and CONADE, is presently the leading ESCO in Chile.

In addition to Dalkia, another foreign company, TBE Chile Asesorías y Representaciones, entered the Chilean marketplace. It is a subsidiary of TBE Energiemanagement GmbH of Austria, which, since 2001, has specialized in contracts to optimize energy efficiency in commercial buildings with central air conditioning.

Surveys of companies and consultants have led to training activities for future ESCOs within the framework of a technical assistance project of the Inter-American Development Bank BID-Fomin with Fundación Chile. It is expected that the number of players in the market will increase to a dozen consulting companies, which could potentially become ESCOs. These companies possess technical knowledge and experience in project development and have already conducted energy audits for commercial and industrial clients. Several took part in a programme to promote market opportunities for investments in "clean energy" (Programa de Promoción de Actividades de Mercado para Energías Limpias), as well as training courses in 2007 by an international ESCO (Econoler), which introduced the operational mechanisms of an ESCO and the way to conduct an economic feasibility study.

Up to now, no equipment suppliers have shown interest in joining this business, either through the creation of their own company, or the purchase of an existing one.

Market and Contract Characteristics

Reviewing the services rendered by companies already established in the local market, we can distinguish the following:

In the case of **TBE Chile**, its main focus is on commercial buildings and industry, offering two approaches: (1) energy performance contracting

(EPC) and (2) procedures for optimizing energy costs (energy audits, etc.). A typical EPC contract by TBE is based on optimizing:
- energy costs through the installation of state-of-the-art technology;
- energy efficiency management in the building or industry, including training the client's staff; and
- the client's rate structure (electricity consumers are divided into two categories for the purpose of pricing. Small consumers with demand of less than 2M\mW, pay for their electricity according to a regulated tariff based upon the marginal costs of electricity generation. "Free," or large consumers [>2M\mW] are free to negotiate price contracts directly with generating companies.)

Investments, financed with their own capital, are limited to a maximum of USD 50,000/contract. Contract duration ranges between two and five years. TBE receives part of the savings, but offers no guarantee as to achieving the predicted savings. To date, TBE has entered into 65 EPCs, most of them in commercial buildings.

Author's note: Without guaranteed performance, these TBE projects do not satisfy the definition of energy performance contracts.

The other type of contract, based on preliminary audits, offers an estimate of potential savings. These contracts include a fixed fee, which is measured in terms of reaching goals without investments by TBE.

On average, TBE clients have obtained 26 percent savings in their electricity bills. Annual sales by TBE total USD 750,000.

Dalkia provides global solutions, including specialized staff to effectively operate client installations. In case the client wishes to outsource facility service and investment, Dalkia offers "chauffage" type contracts in which it establishes the "utility" price, calculates performance, and commits to a savings value, which only varies depending on fuel prices; e.g., steam in cost/ton or compressed air in cost/m^3. "Energy supply" is another type of contract that may or may not include guaranteed savings.

Dalkia offers its services to all segments in the market (except public lighting), but concentrates primarily on the industrial area with approximately 120 thermal power plant energy management contracts, catering to international and domestic clients of the agro-industrial and food industries. Dalkia is the only ESCO with experience in the private hospital sector, and it participates in public-private partnership projects in the public hospital sector (See case study at the end of the Chile report). Its services include financing for the renewal of client equipment, together with the operation of installations, as well as facility management.

The type of project financing is determined on a case-by-case basis, including contracts with total financing of investment through the ESCO's own capital or via co-financing. In addition, Dalkia also operates under customer-financed contracts.

In summary, ESCO activities focus on the commercial sector, buildings and large shopping malls, and on the industrial sector, primarily agricultural and food industries. No energy contracts have been signed with the public sector. Some energy audits have been conducted, but they have not resulted in tenders for contracts.

Estimation of the Potential Market

Some data are available to estimate the market potential. Studies undertaken by the National Energy Commission (Comisión Nacional de Energía CNE) and universities are, however, of limited value when it comes to promoting business plans related to energy efficiency activities. They are only helpful as contributions towards the creation of a general framework for energy policies and to identify market segments with greater potential for energy services.

The following table shows estimates of the expected average annual savings for a 10-year period. The data are based on the study entitled "Estimation of Potential Energy Savings through Energy Efficiency Improvement in Different Consumption Sectors in Chile" made for CNE (2004) and recently updated through research conducted by a project of IDB/ Fundación Chile. The findings are presented below in Table 8-1.

Factors which Promote Market Development

One of the fundamental factors influencing recent market growth is the increase in energy prices. In the case of private regulated clients (under 500 kW) in the centre of Chile, the regulated price increased from USD 32/mWh in 2004 to USD 47/mWh in 2005 and USD 66/mWh in 2006. At present, it may go as high as USD 140/mWh. After a decade of low prices, Chile today has the highest electricity prices in Latin America.

As a consequence, increasing energy efficiency in all sectors has become a great political priority. This is expressed in Programa País de Eficiencia Energética (PPEE), created in 2005, as well as in the establishment of a Policy of Energy Security "Política de Seguridad Energética (PSE)"

Table 8-1. Potential EE Savings

Sector/Sub-sector	Potential EE Increase % per Year between 2004 and 2015	Potential Annual Market of EE USD Million per Year in 2005
Industry		
Pulp and Paper	4.5	21.9
Steel	2.7	4.9
Petrochemicals	2.6	1.3
Cement	1,9	2.4
Sugar	2.7	1,2
Fishing	4.5	3.3
Others	4.1	50.0
Mining		
Copper	0.8	11.5
Nitrates	3.2	2.4
Iron	4.1	2.1
Other Mines	3.2	8.0
Services		
Commercial and Public	2.9	27.7
Residential	0.9	51.2
TOTAL		187.9

Fundación Chile (2007a)

that became official in 2006.

The creation of the PPEE was based on the decision to reinforce the political action around the issue of energy efficiency, and has been based on the principle of public-private partnerships. This initiative is one of the most explicit governmental decisions to face the problem of energy supply, and is an integral part of its energy and environmental policy. Through this initiative, the government seeks to counteract the fact that the national consumption of electricity has almost tripled in the past 15 years, and the intensity of electricity consumption per unit of gross domestic product increased more than 30 percent over the same period. At the same time, the dependency on imported fuels for energy has increased from 18 percent in 1982 to 70 percent in 2006. All this makes the country highly vulnerable to the fluctuation of international prices and creates great uncertainty in power supply. Additionally, the high percentage of hydroelectric generation creates vulnerability associated with local and global climate change.

The strategic plan 2007-2015 outlines specific objectives, including "the development of products and services markets associated with energy efficiency" and "to have ESCOs in the energy services market." One line of action is the development of economic incentives and financing tools, including direct subsidies, tax rebates and/or low interest rate credits as well as the promotion of energy performance contracts.

The following activities, geared towards opening the door for ESCOs in the market include: (1) publication of an "Energy Efficiency Directory, including a compendium of institutions, consulting companies, equipment suppliers, and companies from the energy and industrial sectors, (2) programme of technical seminars describing ESCO international experiences, (3) awards for energy efficiency solutions for the industrial sector, organized jointly with the major Chilean Association of Industry and Commerce, to reward initiatives in industry, including the trade, services and tourism areas.

Chile ratified the Kyoto Protocol in 2002 and fosters projects within the framework of the clean development mechanism (CDM). The field of energy efficiency is one of its priorities.

Technical and Financial Assistance

Some initiatives have already been completed or are underway to support the creation and consolidation of the ESCO market:

(1) the German Technical Cooperation Agency (GTZ) advises PPEE and relevant ministries in the area of energy, on issues related to norms, standards and instruments;

(2) the programme "Programa de Promoción de Actividades de Mercado para Energías Limpias," implemented by Fundación Chile, seeks to support the projects and business of energy service companies; and

(3) the international cooperation project financed by REEEP (Renewable Energy Efficiency Partnership), and implemented by the Berlin Energy Agency (BEA), GTZ and PPEE, called "Energy Services—an Innovative Financing Scheme for Energy Efficiency in Public Buildings in Chile" analyzed the macro-legal conditions and contributed towards the creation of awareness and information in the public sector about different types of contracts.

Since late 2006, the Ministry of Economy has been able to finance up to 70 percent of energy consulting services through the Programa de Preinversión en Eficiencia Energética (PIEE), of the Chilean Development Agency. Those services include financing of the "Project Design Document," for a CDM project. This programme is aimed at companies from all production sectors requiring energy efficiency studies, such as audits, energy efficiency implementation plans, or investment plans to access a source of financing. Companies with annual net sales less than USD 35 million are eligible. Despite the seemingly minor relevance that such an instrument may have for an ESCO, since it is limited to SMEs with small energy consumption, it promotes awareness of this issue at an industry level and encourages the creation of specialized consultants, a significant pre-condition for the growth and strengthening of the ESCO market.

In addition to subsidizing pre-investment studies, a credit line is in place for long-term investments in environmental protection which includes energy efficiency issues. The financing scheme follows the shape of a private bank credit or leasing operation for a maximum amount of USD 1 million with a fixed interest rate slightly below normal credit rates, payback periods between 3 and 12 years and grace periods up to 30 months. The beneficiary company must contribute at least 15 percent of the total investment amount required.

Public Policy Instruments

The "green procurement policy" is envisaged in the PPEE strategic plan. Despite the limited participation of the government sector in the total national energy consumption (less than 1 percent of final consumption), where electricity is predominant, representing nearly 75 percent, its efficient use is of great symbolism for the Chilean society. Through the government procurement agency (Chilecompra) efforts are made to include energy efficiency among the procurement criteria. In addition to this initiative headed by PPEE, it is worth mentioning an economic-environmental promotion policy that establishes "voluntary clean production agreements" between companies, areas of business and government, agreeing on a 3.5 to 7 percent saving goals.

A relevant factor that will facilitate the creation of ESCOs is access to, and almost unlimited availability, of know-how and technology at competitive prices in the international market. This accessibility is the result of the free trade agreements entered into by Chile in recent years with 70 percent of countries in the world.

Main Obstacles to the ESCO Market Potential

According to a survey conducted in 2007 by PPEE in the industrial sector, there is still a surprising deficit of information on the issue of energy efficiency and on the management instruments that exist in the international market. In 2007, around 85 percent of companies across the nation had conducted neither energy audits nor diagnoses, and had no information about their own energy performance in comparison with their competitors. Energy consumption indicators in each productive sector (benchmarking) are practically unknown and are not used as relevant information for competitive strategies. 75 percent of companies that participated in the survey ignore the potential for savings and had no specific goals for reducing energy consumption.

In addition to this disheartening situation, the results of energy audits conducted over the past few years, show that 30 percent of companies had undertaken no energy improvement activity and only 15 percent had completed more than 75 percent of the modifications suggested by the consultants. This poor participation evolved despite the fact that, in most cases, the predicted return on investment was less than one year.

At present, only a very small fraction of major companies with huge energy consumption have an "energy focus man" capable of acting as a competent contact when an ESCO comes forward to present its proposal. For a large part of Chilean industry, dominated by SMEs which represent 70 percent of industries, this issue does not exist as a defined function at an organizational level. Operations and technical managers are closest to this issue, but generally have no decision-making capabilities for this type of investment.

As far as the commercial and construction sectors are concerned, the few preliminary studies conducted to date forecast a potential 2.9 percent annual savings. However, comparative data (benchmarking) is needed, and there are few technical standards regarding energy efficiency.

Potential energy savings in public buildings (schools, hospitals, local authority buildings, etc.) is widely unknown and, therefore, no advantage is taken of it. There is no registry of energy consumption in public buildings, or relevant information to make comparisons. If an ESCO is to be awarded a contract in a public building, it must necessarily go through public tenders and a process to comply with budgetary law, which is presently slow and cumbersome. There are no standardized contract models to facilitate this process.

Financing Concerns

Apart from the lack of useful information, there is another fiscal barrier: exemption of the 19 percent VAT payment for energy costs and related services, which presently benefits public buildings such as hospitals, town halls, schools and pension fund administrators. This fiscal benefit, however, is not passed through to an ESCO. Thus, their energy services for those clients must first of all compensate this 19 percent and on top of this, try to be profitable.

For an ESCO in the international market, the annual minimum energy cost that the company takes as its basis to select a potential client is typically about USD 500,000. For an ESCO in Chile, that amount is USD 100,000, thus restricting the investment margin to make efficiency savings profitable. In addition, the payback periods demanded by Chilean companies are very short—one to three years—hence, further limiting the scope of action of an ESCO.

Transaction costs in Chile are high and there is usually a one-year delay before a contract is signed. That "waiting time" considerably hinders the entrance of new ESCO players, especially small companies. Therefore, it can be said that present market conditions favour the branches of major international companies that can afford a long-term strategic vision to penetrate the market.

Lack of financial mechanisms for energy efficiency projects has made their start-up more difficult. This is not related to the scarcity of competitive commercial banks, because according to the World Bank classification, the Chilean financial market is one of the most developed in the world (with 19 national and 6 foreign banks, as well as the Central Bank). The problem is primarily credit conditions that cannot be accommodated in the financial arrangements of EE projects. Generally, it is short-term credits, for a maximum period of 36 months, that are granted. Different conditions apply to interest rates (basic amount fixed by the Central Bank plus supplement) and the establishment of the supplement amount for each credit depends on the estimated magnitude of risk of the respective project. In January 2008, the annual interest rate of different Chilean banks fluctuated between 8 and 33 percent, depending on the amount of the credit and maturity terms.

Supply Governance

A significant economic aspect is the concentration of power generation and distribution in a few private companies. This is the result of the political reform of the energy sector introduced in the 1980s. That reform led to the granting of supply area concessions to a distribution company that was thus able to benefit from monopolistic conditions. At the same time, the sphere of state intervention in the energy sector was restricted exclusively to supervising and fixing prices for regulated clients. These and other aspects of energy legislation have hindered the entry of further economic actors to this area, restricting competition between companies in the provision of services, thus limiting the offer of ESCOs and preventing them from supplying power.

Future Outlook

The market is expected to grow moderately in the next three to five years. Experts agree on forecasting an economically attractive savings potential in industrial, commercial and public sectors. Higher energy prices are perceived as a major challenge for the competitiveness of the Chilean economy. Other factors affecting the ESCO business potential are related to the idiosyncrasy of Chilean entrepreneurs who—regardless of being open to the world—are reticent to change and may be characterized by conservative management and lack of enthusiasm for new organization styles and innovative financing tools. This conservative nature is expressed primarily in their refusal to delegate functions and decision-making to third parties. Rather than the possibility of being able to resolve a budget issue while concentrating on its core business, they prefer to solve the energy challenge by hiring an additional employee.

An ESCO representative described that attitude by means of the client's tacit question, "Where is the problem?" and not "Where is the opportunity?" The cost of convincing a company to leave a service in the hands of third party experts is high for an ESCO. This attitude prevails in both the private and public sector, where the issue of outsourcing is viewed negatively as a need to lay off employees. This, in turn, leads to labour and union problems.

The task of persuading their clients will be ongoing for ESCOs. There is no ESCO association in Chile and the powerful industry and commerce associations have so far failed to promote and improve this new service sector.

The lack of knowledge and accreditation of measurement and verification (M&V) instruments and procedures favours the doubtful attitude of potential clients about the novel offer of this type of services. Considering these factors, real prospects for market growth in the next few years are not very encouraging.

CASE STUDY:

Public-private Partnership Project

The PPEE's first experience with the introduction of "energy contracting" began in 2006 with a pilot public-private partnership project (PPP) focused on the "Improvement of energy efficiency in public hospitals through the introduction of third-party financing (energy contracting). Headed by PPEE, the project brings together the Ministry of Health and its respective services, a representative public hospital, DALKIA and GTZ. The decision to use the health system to exemplify the introduction of energy performance contracting in the public sector was not taken at random. According to experiences in European public hospitals, it is possible to achieve 25 to 40 percent savings.

One of the biggest obstacles is the lack of detailed information on energy consumption. Furthermore, data registered to date reveal that regulations already in existence regarding energy use in hospitals are unknown and/or not complied with. In addition, available registries are insufficient to develop the energy consumption baseline. Consequently, Dalkia will have to invest heavily in measurement and monitoring equipment during one and a half years for the collection of the relevant data in order to have a sound basis for the implementation of further energy performance contracts. In this phase, it guarantees five percent savings on the energy consumed by the hospital.

Human resources are an important issue in the public sector. In the case of the PPP, Dalkia offered the incorporation of the hospital staff into its work team. However, the limited capabilities and scarce technical knowledge of the hospital staff, as well as the salary and pension benefits of the public hospital sector, prevented such incorporation. In order to alleviate this difficult issue, GTZ has suggested the creation of a qualification programme to facilitate the relocation of staff that cannot be hired by the ESCO.

Based on the results of the pilot project presently underway, a bidding process or model contract will be presented to the Ministry of Health, so that it may use it with energy contracting providers. In order to consolidate the method, different training activities will be conducted with the hospital staff and Ministry of Health officials, and manuals containing the contract principals will also be produced.

URUGUAY

Mr. Marcelo Gonzalez provides an exceptionally comprehensive look at a country, which pays little attention to energy efficiency or its potential benefits to the economy or the environment. Within this context he offers a very perceptive view of conditions for ESCO development. To understand the evolution of the ESCO concept in Uruguay it is important to make a quick overview of the energy sector in the country. The most important aspect to mention is that Uruguay is 100 percent dependent upon imported oil.

In the late 1970s, the Ministry of Industry, Energy and Mining created the Grupo Racionalizador de Energía Industrial (GREI) to provide government paid audits and project design recommendations to industrial energy consumers. The program used energy specialists from academic institutions but the programme had little credibility with the facility managers. So it had very limited impact and almost no records remain of these experiences.

Through the following years, there have been few government-driven efforts to improve energy efficiency in Uruguay and they have generally not yielded significant results. As a precursor to the ESCO industry, it is worth mentioning a private sector driven effort exploiting the opportunity created by a price difference between fuel oil and wood. A couple of Uruguayan industrial boiler manufacturers replaced oil-fired boilers with wood-burning boilers, putting the operation of the new installations on a performance contracting basis.

The introduction of natural gas as a new primary energy source in Uruguay energy matrix dates back to the mid-1990s, with the issuance of the concessions for the natural gas distributors and initial work on the key natural gas interconnections with Argentina. Natural gas is far more expensive in Uruguay than the liquid fuels that it is supposed to replace, which reduces economic incentives for energy efficiency projects.

During the same period, UTE (Uruguay state-owned utility) introduced time-of-use tariffs to alter the load curve and reduce peak demand through an economic signal, with the relationship between the peak and off-peak levels about 4:1. The utility also established a new minimum admissible power factor limit, with stronger economic penalties, to reduce reactive energy in the transmission and distribution networks. In 1999 a programme was launched with sales of commercially available products based on technologies of interest (hot water heaters, heat storing radiators, electric boilers, heat pumps) to specific market segments. UTE followed this in 2000 with a new programme called "Superplan." It was a financing program allowing customers to acquire electricity-consuming household appliances, with peso-denominated financing, at a relatively low interest rate.

In August 2004, Uruguay signed with Global Environment Facility (GEF), a grant agreement with the World Bank, for the development of the Uruguay Energy Efficiency Project (UEEP), with the objective to increase consumer demand for, and competitive supply of, energy efficient goods and services. The project addresses barriers to energy efficiency investments, including the promotion of ESCOs to deliver energy efficiency services. Within UTE, project funds will leverage UTE resources to create an Energy Savings Unit (ESU) that will operate on ESCO principles. In parallel, technical assistance will be provided to develop contractual instruments for energy performance contracting and to train companies interested in operating as ESCOs.

Current ESCO Industry and Market

Until the beginning of the year 2000, the knowledge of the ESCO concept was limited to a small number of professionals interested in the energy sector and only as an academic matter. A number of small and medium-sized engineering firms participating in the energy market, began covering many of the tasks a "real" ESCOs would perform, but did not guarantee the results.

We can differentiate the following types of firms: energy advisors, engineering consultant firms, quasi-ESCO firms and UTE-USCO.

Energy Advisors

In this category, are electrical installation firms and small-sized engineering consultant firms, which advise customers as to the best electricity tariff to use according to the way the electricity is consumed, and to reduce the economic penalties for the reactive energy consumed. In the latter case, some firms finance the installation of capacitors for power factor correction (because they are inexpensive and have a short payback period) with the implementation of shared-savings contracts. But in general they have limited capitalization, short corporate histories and limited sophistication in terms of mechanisms for raising capital and obtaining project financing.

Engineering Consulting Firm
This category includes medium-sized engineering firms that have extensive experience in engineering design, project implementation, follow up capabilities on construction and commissioning, supervision and the financial side of the performance contracting business. Their knowledge of the ESCO concept is limited. They have not yet embarked on the financing of projects because the usual project size is rather large relative to the size of the firms.

Quasi-ESCO Firm
A couple of engineering firms, which might be categorized as quasi-ESCOs, are almost working as ESCOs. Similar to the energy advisors, they have limited capitalization for financing projects and for arranging third party financing schemes.

UTE-USCO
Under Uruguay Energy Efficiency Project Component 2, Utility-Based Energy Efficiency Services, UTE is designed to assist in the creation and operation of an Energy Savings Unit (UTE-ESU). New business activities for UTE include energy efficiency services, especially demand-side management. It would offer services to UTE, and through UTE, to its clients; e.g., financing more efficient equipment through payments on utility bills and directly to business clients. Since part of the engineering expertise needed to develop an ESCO project is not currently available inside UTE, the use of private energy service companies appears to be an attractive alternative and could speed up the beginning of operations.

UTE-USCO (EFICENER) receives part of the GEF grant for financing the implementation of energy efficiency projects on a performance-contracting basis.

Econoler International has been awarded a contract with UTE for the start up of the UTE-USCO on an ESCO basis, through transfer of know-how by the practical implementation of energy efficiency projects.

Uruguay Energy Efficiency Project—List of ESCOs
As a component of the Uruguay Energy Efficiency Project, the National Energy and Nuclear Technology Office (DNETN), is being used to promote the development of the ESCO industry. One of the actions taken has been to create a registry of ESCOs in Uruguay. Anyone interested in

becoming an ESCO simply has to fill out an internet application form at the internet website (www.eficienciaenergetica.gu.uy).

Financing

Until recently there has been little financing of energy efficiency projects by the ESCO sector. Most energy efficiency projects have been financed by the client's own money. In the few cases where the ESCO has financed the project directly, financing came from the energy firm's capital.

With the implementation of the Uruguay Energy Efficiency Project, two new financial sources are becoming available, GEF funds directly used by UTE-USCO through performance contracting with clients, and an Energy Efficiency Fund established with part of the GEF grant. Basically, this new fund will operate as a guarantee fund.

The following chart describes the financing scheme in process of implementation.

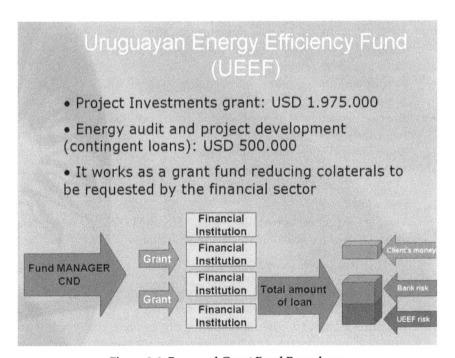

Figure 8-1. Proposed Grant Fund Procedures

Contracts

When financing with an ESCO's own money is included in the project, a formal contract document is signed between the parties, establishing different aspects of the relationship during the contract duration, including the amount of periodic payments, contingencies, etc.

In such cases where there is no financing on the part of the ESCO, to date very little formality has been required. Frequently, no document is signed at all. One important reason for this behaviour is the relative small size of the Uruguayan market, where the different actors (consultants, equipment suppliers, industry owners) in the industry are well known to each other.

In the case of UTE-USCO, performance contracts are now being prepared and adapted to the Uruguayan conditions. These contracts are usually "first-out" contracts or shared-savings contracts, depending on the type of client, public or private.

ESCO Projects

To illustrate the type of projects private ESCOs are proposing, Table 8-2 lists a recent call for presentation of energy efficiency projects to be partially financed by the Uruguay Energy Efficiency Project.

UTE-USCO is now working on the following projects:
- Public lighting;
- Retrofit of lighting and ambient conditioning installations in public buildings (offices, hospitals, university);
- Energy efficiency improvements of industrial auxiliary processes (compressed air, chillers, boilers, etc.);
- Solar home systems for rural low income population; and
- Energy efficiency in poor neighborhoods of Montevideo.

Barriers to ESCOs

During the preparation of the Uruguay Energy Efficiency Project the following barriers for the implementation of energy efficiency projects were identified:

> *Lack of demand for energy efficient goods and services.* Public knowledge about the financial and economic benefits of energy efficiency measures is limited. The market does not provide customers with information on potential cost savings, energy savings and emission reductions. For example, vendors of appliances and equipment pro-

South America

Table 8-2. Typical ESCO Projects

Project #	Client Name	Activity Sector	Saved Energy Usage	Energy Source	Location
1	IMM	Municipality	Public lighting	Electricity	Montevideo
2	LANERA PIEDRA ALTA S.A.	Wool industry	Heating, vapour circuit	Wood	Florida
3	IMP	Municipality	Public lighting	Electricity	Paysandú
4	CURTIFRANCE S.A.	Leather industry	Heating, hot water	Fuel Oil	Montevideo
5	CRUFI HELADOS S.A.	Food industry	Heating, cooling cycle	Fuel Oil	Paysandú
6	PILI S.A.	Food industry	Heating, hot water	Natural Gas	Paysandú
7	LATU	Laboratory	Heating, hot water	Fuel Oil	Montevideo
8	IMM	Municipality	Heating, hot water	Electricity	Montevideo
9	PROLESA	Agriculture/milk	Heating, renewable	Electricity	Florida, San José, Canelones
10	FLORIDA	Wool industry	Peak shaving	Wood	Florida
11	CRISTALPET	Plastics	Compressed air	Electricity	Montevideo
12	PORTONES SHOPPING	Shopping centre	Air conditioning	Electricity	Montevideo
13	FRIMACAR	Food industry	Heating, vapour circuit	Natural Gas	Canelones
14	PAMER S.A.	Paper industry	Heating, combustion	Wood	Montevideo
15	MOTOCICLO	Metallurgic industry	Heating, vapour circuit	Fuel Oil	Montevideo
16	BSE	Public hospital	Lighting	Electricity	Montevideo
17	FNC	Food industry	Motors	Electricity	Montevideo
18	BADER	Wool industry	Motors	Electricity	San José
19	MOVIECENTER	Cinemas	Air conditioning	Electricity	Montevideo
20	PUNTA CARRETAS SHOPPING	Shopping centre	Air conditioning	Electricity	Montevideo
21	COFRISA	Food industry	Cooling	Electricity	Montevideo
22	HOSPITAL EVANGÉLICO	Private hospital	Heating	Fuel Oil	Montevideo
23	HOSPITAL ITALINANO	Private hospital	Heating, renewable	Fuel Oil	Montevideo

vide little information about annual operating costs or energy use that would help purchasers to identify the most efficient units. Energy efficiency standards are lacking, as are testing, certification and labeling of electrical equipment. Industry associations are not actively promoting energy efficiency.

Businesses in Uruguay, as elsewhere, tend to invest in capital equipment to increase output rather than reduce costs.

Limited capacity and know-how among key stakeholders. Information on energy efficiency measures or the ways to structure, finance, and operate ESCO-based saving initiatives is scarce in Uruguay. Effective mechanisms for disseminating information to users, policy makers, and regulators have not been developed.

Lack of project development and investment financing. As in other countries, the lack of experience in commercial financing for such projects poses the most important obstacle. In addition, the market infrastructure needed to provide technical assistance in project design, financing, implementation or verification does not exist. Further, the few projects undertaken on a performance contracting basis have not received commercial financing. In the case of the existing ESCOs, company principals have used personal loans or company resources to secure funds.

Availability of credit from the banking system. Energy efficiency projects will be difficult to finance until commercial banks gain hands-on experience. Credit from the banking sector is now reported to be limited and costly, mainly available in US dollars to firms that have income in foreign exchange and/or the ability to provide substantial guarantees. Since energy efficiency savings are earned in local currency, financing is difficult for energy efficiency investments.

Uncertainty about economic growth. From 1991 to 1998, Uruguay's economy grew at an annual average of 4 percent. However, GDP contracted by 2.8 percent in 1999, 1.4 percent in 2000, 3.4 percent in 2001 and 10.8 percent in 2002. Growth resumed in 2004, but the recovery has been slow. Consequently, the willingness and ability of companies to invest in energy efficiency has been very limited.

Relative energy prices. Many of the investment opportunities identified by project studies for efficiency improvement are contingent upon a switch by industrial and commercial enterprises to natural gas. However, if prices fail to reflect economic costs, fuel-switching with attendant opportunities for energy efficiency may not happen on a large scale.

With the experience gained by UTE-USCO, additional barriers have been identified: They include:

Public procurement rules. The different steps in public procurement rules required for guaranteeing the fairness and transparency of the acquisition and contracting processes impose unacceptable intervals of delay for the private sector. At present, it is very difficult for a public office to enter into a contract with a private ESCO through a bidding process if the ESCO wants to implement a shared-savings performance contract.

Budgetary restrictions. For public offices, it is difficult to receive the benefits of a reduction of energy consumption made through an energy efficiency project in its own premises. These offices receive very little money that can be invested in such efforts.

Level of comfort. Working conditions in some cases do not comply with the minimum recommended comfort standards. This is frequently due to a lack of maintenance or budget restrictions. To implement an energy efficiency project in this type of situation, complying with the standards is required, which imposes very long payback periods (more than 5 years) and may actually increase energy consumption in the installation.

Enabling Factors for the Development of the ESCO Industry

New legislation, an Energy Efficiency Law, has been under preparation. It is of paramount importance to make explicit for all the stakeholders the involvement of the government in the concept of energy efficiency, and the value of energy efficiency to Uruguayan society. This legislation must define clear goals to achieve in the short, medium and long term, the responsibilities of each actor in the energy market and the financing sources to be used.

The Uruguayan Energy Efficiency Fund is another important positive enabling factor that leverages the working capacity of the existing private ESCOs' funds. This will, in turn, allow other potentially interested firms to perceive the possible benefit of this new kind of business.

The ESCO Market in Uruguay in 3 to 5 years

It will still take a couple of years for UTE-USCO to have sufficient project experience to operate as a true ESCO. After this start-up effort, UTE will have to decide the best way to gain all the benefits of this kind of endeavour. Options include leaving UTE-USCO as an internal unit of UTE, or give UTE-USCO a different status outside UTE.

The implementation of the Energy Efficiency Fund will enable the private sector to establish new ESCOs, bringing comprehensive solutions for energy efficiency projects. The Uruguay Energy Efficiency Project has set what it considers a reasonable objective of identifying ten effectively working ESCOs at the end of the project.

BRAZIL: CHALLENGES OF A FAVORABLE MOMENT

The ESCO business in Brazil dates from the mid-1990s, though some engineering firms had begun to carry out energy audits and offer specialized energy rationalization services in the early 1980s. The first seminar on the concept of energy performance contracting was held in 1995 and in 1997 the Brazilian ESCO Association (ABESCO) was founded.

Since then the growth of the sector has been substantial, though any estimate must be very imprecise because there has been no attempt to systematically survey the sector recently. There have been periods of reverses as well as advances. At the same time, there has been a development of capabilities which continues today. The ESCO sector has matured more than the increase in volume of business alone would suggest. Alan Poole, who lived in Brazil from 1981 to 2008, has been heavily involved in energy efficiency and ESCO development during that time, and offers a very complete review of this maturing industry. His discussion of the impact of political shifts on the industry's development is of particular interest.

Many of the historic barriers continue to limit the growth of the ESCO business. There have been some recent initiatives from the government, however, which could stimulate both faster growth in the volume of business and the capabilities of the firms. This government action to explicitly promote the ESCO sector is, in itself, something of a novelty. Previously there had been almost no official encouragement. The one partial exception involves utility DSM programs.

The author uses the term ESCO according to common Brazilian usage—that is, including conventional fixed fee contracting of engineering services. When appropriate, specific reference will be made to energy performance contracting (EPC). Only a few firms in Brazil would qualify as an ESCO in the strict definition of the term as an energy efficiency performance contractor or as a "full service ESCO."

Nevertheless, these firms do offer specialized energy efficiency (EE) services, are prepared to guarantee a level of savings and to verify their performance using M&V procedures. These are features which distinguish them from general engineering consultant firms.

ESCO Development

As in a number of other countries, the ESCO business in Brazil had its beginnings in the performance of energy audits for government programs.

Energy efficiency first became prominent in Brazil in the early 1980s, as a result of the oil price shock. A government-sponsored program known as CONSERVE was started to encourage the conservation of petroleum derivatives and fuel substitution in Brazilian industry. The program sponsored nearly 600 audits, reached accords for oil savings targets with major oil consuming sectors, and offered low-interest loans for project implementation. This program was relatively successful; however, fuel substitution predominated very much over efficiency improvements. While many housekeeping measures and simple retrofits were implemented, very few industries made major efficiency investments. The sharp decline in oil prices after 1986 led to a drop in interest in the rationalization of fuel use, which continued until recently.

The national program to promote the efficiency of electricity use—PROCEL—was created in 1985. In its first years this program funded many audits of electricity use, but few investments in actual retrofits occurred. In the early 1990s, hyperinflation and economic turbulence, together with the political decline of PROCEL, brought EE activities almost to a standstill. Most firms in the business moved to other consulting services and/or drastically downsized. This was a difficult period for engineering services in general.

In 1993 the power sector regulator required medium and high voltage consumers to achieve a minimum power factor in order to reduce grid losses due to reactive power. Significant fines for non-compliance were imposed. This rule opened a niche market, which exists today. It also helped "open the doors" of potential clients. While correcting the power factor, ESCOs had an opportunity to identify EE measures and present them to their clients.

This specific stimulus was amplified by the broader economic changes which began in 1994: inflation fell dramatically, the price of electricity rose and became more stable, duties fell on imported goods, and information technology became cheaper.

At the same time, the national PROCEL program was re-activated. While it did almost nothing to directly promote the ESCO concept, the program did heighten awareness among consumers of the benefits of EE measures for electricity.

This period saw the beginning of the definition of the ESCO business and the public discussion of energy performance contracts. The Brazilian ESCO association (ABESCO) was founded in 1997 with 15 members. A survey of the ESCO sector at that time estimated that R$ 16-17 million (USD 16-17 million) of projects were implemented in 1996.

An important new market was opened by legislation in 2000 regulating utilities' DSM programs. Under this DSM policy—known as the Energy Efficiency Program or EEP (see box)—utilities must spend a percentage of their revenue on energy efficiency projects with consumers under the oversight of the regulator of the power sector, ANEEL. The law allowed utilities to invest up to half their EE resources in projects in which they could recover the cost of the investment. The projects were structured as "performance contracts" between the utility and the beneficiary, where payments were included in the utility bill (and could not exceed the monthly savings).

Many utilities soon opted to contract ESCOs to execute these projects. It is estimated that in 2002 about 117 projects worth R$ 23.5 million (USD 8.7 million at the time) had been contracted to ESCOs by utilities. This segment rapidly became the biggest single market for ESCOs in Brazil and has been an important source of revenue for many ESCOs. It must be emphasized, however, that ESCO contracts with the utilities used standard cost-plus engineering services, not performance contracts, for this work.

The year 2000 also saw the publication of a presidential decree, which required that all federal buildings reduce their electricity consumption by

20 percent within two years. This raised hopes that something like the federal building programs of the US and Canada could emerge. Such programs had played a big role in consolidating the ESCO business in these countries. Unfortunately, the legal issues about tendering performance contracts for government buildings were not resolved, so little happened and the public building market has remained largely closed for ESCOs.

The Utilities' Energy Efficiency Program Regulated by ANEEL

Since 1998, it has been mandatory for electricity distribution utilities in Brazil to invest in energy efficiency. The model that was adopted for the Energy Efficiency Program (EEP), as it is called, requires distribution companies to propose and carry out energy efficiency projects with the oversight and regulation of ANEEL (the regulator of the power sector). The program and its experience over the years is attracting increasing interest in diverse countries with emerging economies, which are trying to strengthen their energy efficiency policies and are seeking instruments for this purpose.

The resources for EEP are defined as a percentage of the Net Operating Income (NOI) of the utility. The share allocated to energy efficiency has varied over the years and is currently at 0.5 percent of the NOI. This currently generates more than R$300 million (USD 200 million) per year for EE projects. A similar amount of resources goes to R&D programs, which are also supported by this "wire-charge."

The EEP is the largest source of resources for financing energy efficiency initiatives in Brazil. Since its inception in 1998 about R$1.8 billion (USD 1.2 billion) has been invested in energy efficiency projects through the program. It has been subjected to many changes since its creation.

Law 9991 in 2000 consolidated the regulatory framework for the application of resources resulting from the utility "wire-charge" for energy efficiency, after operating on an ad hoc basis for several years. The program had been created in order to assure that the "public interest" in energy efficiency investments would be maintained after distribution utilities were privatized. It was then extended to all distribution utilities, regardless of whether they had been privatized.

The most important innovation, from the perspective of the ESCO business, was to allow utilities to invest up to 50 percent of the resources allocated to EE in projects where they could recover the cost of the investment. Previously all projects had been on a grant basis. The resources recovered could be dedicated to future EE projects in general. The projects were structured as "performance contracts" between the utility and the beneficiary. The utilities then hired ESCOs to implement many of these projects with standard cost-plus engineering contracts. This rapidly became the single most important source of revenues for Brazilian ESCOs.

(Continued)

> In 2005 there were again extensive changes. Various types of projects in which there had been abuses were prohibited. A large new category of projects was created to assist low income consumers. At least 50 percent of a utility's EE resources must go to projects in this category. In practice more than 60 percent have gone this way. This and other changes meant there was a severe contraction in the resources available for the kinds of projects that ESCOs might execute. Also, about this time the utilities were obliged to reinvest all the income from their "performance contracts" into this same category of project. This new obligation had the immediate effect of drying up utility interest in this kind of project.
>
> Despite the reforms over time, there has been concern about the efficacy of these expenditures, especially in terms of increasing general levels of the efficiency of energy use or market transformation, as well as a lack of adequate evaluation. Consequently, ANEEL began an analysis of possible reforms in 2007 with support from the World Bank. New rules and a manual to orient the utilities were published in February 2008 by ANEEL. The new regulations greatly simplify the presentation and approval of most projects while placing more emphasis on adequate verification of results. This emphasis may result in more training of M&V methodologies.
>
> Several new categories of projects are allowed. One category permits, for the first time, projects which are not proposed and managed by the utilities. In principle, the new regime for the EEP should increase opportunities for ESCOs. The counter-productive obligation to reinvest income from projects with performance contracts, which was described above, has been revoked. This has removed a big disincentive to utilities using performance contracts and should have favourable impacts for ESCOs in the short term.

Finally, 2000 was also the year that a large program—named RELUZ—was launched specifically to finance energy efficiency in public lighting. During the period 2002-2004 the total financing was about R$ 660 million (USD 220 million at the time). By the end of 2004, contracts had been signed with more than 1400 municipalities, roughly a third of the total in Brazil. It was expected that three million lighting points would be substituted as a result of this financing—out of a targeted total of about 12 million. Only a very few ESCOs participated in this market segment. In more recent years the rhythm of expenditure of RELUZ has fallen dramatically, mainly due to new difficulties in the project approval process.

One curious thing stands out about spending priorities for energy efficiency (EE) in the early 2000s. Besides the resources from RELUZ, public lighting received the lion's share of the utilities' EEP resources as well—more than half. Overall, about three fourths of total spending on

EE in the period went to the niche of public lighting retrofits—responsible for about three percent of electricity consumption. The situation has changed since 2005. From that year, public lighting projects were prohibited in the EEP and RELUZ has since had problems approving projects. Equipment manufacturers, having geared up for the stream of expected orders of public lighting lamps and fixtures are now alleged to have idle capacity.

The flurry of activity in 2000 may have been spurred by government unease about the rate of expansion of electricity supply. In 2001 subnormal rainfall exacerbated an already precarious supply-demand balance and led to electricity rationing from June to the end of February 2002 which required cuts of 20 percent by most consumers. However, this crisis did little to develop the ESCO market.

A big reason was that the rationing was announced quite suddenly. As late as March 2001, the government and the ONS (operator of the national grid system) denied that there would be a supply problem. Rationing then began on June 1 of that year. All this left very little time to develop and negotiate projects with consumers, especially anything resembling a performance contract. The full reduction in consumption had to be achieved starting in the first month of the rationing, and most consumers opted for "self-medication." The main exception was a rush to buy backup generation sets, which briefly benefited some ESCOs.

During the rationing crisis, the government showed considerable interest in developing a longer term "market oriented" energy efficiency policy, with the promotion of ESCOs' services as a key element. This was the first time that government energy policy had explicitly recognized the potential importance of EE services. The energy glut and the entry of a new Federal Administration in January 2003 led to the abandonment of this promising initiative. In early 2003 the new energy minister famously stated, "Energy conservation is no longer fashionable."

However, while there was a glut of electricity for several years after the rationing crisis, prices for "captive" consumers continued to increase substantially above inflation, improving the economics of efficiency projects. Another favourable factor was the expansion of the distribution network of natural gas. Consumers switching to this fuel represented an opportunity to identify broader energy rationalization measures, analogous to what had happened earlier with "power factor" correction. Unfortunately, very few ESCOs were prepared to exploit this opportunity.

The main factor sustaining the ESCO sector during this period was the utility wire-charge program described above. One can, therefore, imagine the near panic which gripped the sector when the government proposed in mid-2005 to transfer this resource from energy efficiency to projects to assist low income residential consumers. After much lobbying only a part of the resources were re-allocated.

On a more positive note, in May 2006, the National Bank for Economic and Social Development (BNDES) approved an innovative new credit line, called PROESCO (see box opposite). It was designed to address the problem of guarantees for loans to EE projects—a long-standing problem. The BNDES is the ultimate source for almost all medium-term commercial bank debt financing in Brazil.

The post-rationing glut of electricity is now a distant memory. The Brazilian economy has begun to grow faster than in the previous two decades and there are some worries about the future supply/demand balance for electricity.

At the same time, awareness of the issue of climate change has increased and international studies, which show efficiency as a relatively low cost approach to reduce green house gas emissions, have been highlighted in the Brazilian media. All this has made energy efficiency "fashionable" again. Various initiatives are beginning to be implemented or developed.

Characteristics of the ESCO Business in Brazil Today

The ESCO industry in Brazil today is quite diverse both in terms of the types of ESCOs in the market and the types of projects which are implemented. However, any attempt to characterize the ESCO sector is severely hampered by the lack of systematic surveys.

The most complete survey was performed in 1997, when the sector was still incipient. Partial surveys were carried out in late 2003 and early 2005. These later surveys, while useful for describing many aspects of the industry, were not designed to estimate the volume of business in terms of the number and value of projects. Thus, while one can say that the industry has expanded substantially from the USD 16 million estimated in 1996, there is little basis on which to say how much. An educated guess is that the total value of new projects begun in the past year was about USD 60 million or a bit more.

Today ABESCO has more than 50 members, most of whom are firms providing specialized energy efficiency services in Brazil. There are also some firms offering energy efficiency services, who are not yet

> ### THE PROESCO RISK-SHARING CREDIT WINDOW FOR EE PROJECTS
>
> In May, 2006, the BNDES launched a new variant of an existing credit line for EE, called PROESCO. The credit line is offered via commercial banks, who serve as intermediating financial agents (IF). Some of the key innovations are:
> - The BNDES shares up to 80% of the credit risk of the loan with the IF. Normally the IF is responsible for all the credit risk of the operation.
> - The BNDES shares proportionately in the authorized remuneration of the loan, beyond its normal administration fee.
> - The BNDES relaxes its normal requirements for collateral, emphasizes the quality of project receivables and establishes guidelines for the IF to do so as well.
>
> An agreement must be signed between the BNDES and any IF willing to accept the terms of PROESCO. So far adherence by IFs has been slow. The second bank to sign the necessary agreement did so in early 2008. Both banks are large.
>
> ESCOs would have access to loans under the following terms:
> *Long-term Interest Rate (TJLP) + Administrative spread for BNDES + Administrative spread of the IF + Spread for risk (divided between BNDES & IF)*
>
> Under current conditions, with existing nominal interest rates and spreads, the values would be:
> 6.25% + 1.0% + 1.0% + 3.0% = 11.25%
>
> The small number of participating banks has slowed down the emergence of proposals. Development of the new guidelines of PROESCO within IFs' own organizations also presents challenges.
>
> Despite these difficulties, it is claimed that about 40 proposals are now under consideration or have been approved. After start-up problems, PROESCO may be poised to serve as a catalyst for considerable growth in the ESCO market.

members of ABESCO.

Most ESCOs are independent engineering consultant firms. However, there are a few which are subsidiaries of larger companies which manufacture equipment, such as Johnson Controls, or are energy utilities; e.g., Light ESCO, Efficientia and Iqara Energy Services. Several of these subsidiaries entered the market fairly recently, or strengthened their operations.

The majority of ESCOs are quite small and sell their services within only one of the five regions of Brazil. The surveys carried out in 2003 and 2005 found that roughly one half had less than 10 employees (12 employees in the 2005 survey). At the same time 40 to 55 percent of ESCOs had income from EE below R$1 million (worth about USD 325,000 in 2003 and USD 410,000, in 2005, respectively). However, there are some medium sized firms as well as the subsidiaries of larger companies mentioned above.

Few ESCOs restrict the scope of their services to energy efficiency (EE) measures alone. About 30 percent of those surveyed earned less than one-half of their income from EE measures, while 40 to 50 percent earned more than three-fourths from this source. Common additional services provided include: renegotiation of tariffs with electric utilities, power factor correction, electrical wiring, reduction of harmonics, stand-by generators, distributed power, facility maintenance and rationalization of water use.

With regard to energy efficiency as such, there is an overwhelming dominance of measures to reduce or optimize electricity use. The frequency of projects to reduce fuel use is much lower, though about one-half of ESCOs claim to provide services in this area. One reason for this dominance is the very small consumption of fuels in Brazilian buildings, especially if cooking is excluded. Space heating, for example, is almost non-existent.

Commercial buildings are the largest market segment for ESCOs, followed fairly closely by projects in industry. The residential sector has attracted little interest, while the public sector has been almost absent as a market, due to difficulties in procurement. There have been only two relatively small examples of procurement of an EPC in the public sector.

Unfortunately, there is relatively little information available about the characteristics of these performance contracts. The most common basis for remunerating performance contracts is the savings effectively achieved in monetary terms (Reals). However, savings in terms of physical units; e.g., kWh, are also common as a basis. In general one can say that Brazilian ESCOs are quite flexible in the methodology used to calculate remuneration.

The most common approach to finance projects is for the ESCO to provide the financing, especially in the commercial segment. However, it is almost as common for the client to provide the financial resources, usually from his own cash flow, but occasionally by taking out a loan.

Barriers and Challenges

When ESCOs provide the financing for a project it is almost always with their own capital. Loans to ESCOs to finance EE projects have been very rare. The lack of access to third party financing has historically been a key barrier to the expansion of energy efficiency services.

Although the ESCO industry and energy performance contracting have expanded substantially over the years, the volume of business is still well below the potential for this market. The barriers include the "usual suspects," but there are particularities which are specific to Brazil. All of the following inhibiting factors are present in the Brazilian market.

a. Awareness and perceptions of the potential client;
b. Difficulties in the decision-making and/or procurement process of the potential client;
c. Distortions in energy (or other) pricing which effect the economic viability of projects;
d. Credibility of ESCOs and of the performance contract mechanism; and
e. Financing of projects, especially access to third-party financing.

a. Client Awareness and Perceptions

This set of barriers was ranked very highly by the ESCOs. The table below shows some differences in the relative weights attributed by the two surveys by ABESCO and GERBI (Canada). The biggest discrepancies occurred with respect to problems (1) and (3) in the table. In (1) "Poor customer understanding of potential benefits," the importance attributed by ESCOs in the ABESCO survey was much less than in the GERBI survey—6th versus 3rd out of a universe of eight ranked factors in both surveys.

Table 8-1. Private Client Perceptions as ESCO Barriers

	Problem GERBI 2003	Ranking * ABESCO 2005
1) Poor customer understanding of potential benefits	3	6
2) Customer believes he can do the work himself	5	2
3) People responsible for O&M feel threatened by service provider	6	5
4) Low priority of energy efficiency improvements for customer	2	3

*Ranking was out of a total of eight possible factors which were considered.

Conversely, the ABESCO survey ranked the problem "Customer believes he can do the work himself" much higher as an issue which could derail a negotiation. Perhaps the latter is only a somewhat more sophisticated subset of the first problem "poor customer understanding of potential benefits." It is poor understanding of the value added by the ESCO and its guarantees which is the impediment.

b.   Difficulties in the Decision-Making or Procurement Process of the Potential Client

The "complexity of the potential customer's decision-making process" was considered to be a significant problem, arguably the third most important overall in both surveys. It should be born in mind that the emphasis in both surveys was on negotiations with the private sector.

However, interest in the subject has returned at all three levels of government. Measures to clarify and reduce the legal uncertainties are expected soon. This could open a substantial new market for ESCOs, though questions of how to manage such a program and provide adequate incentives still need to be addressed.

Although performance contracting with ESCOs has been almost nonexistent, it should be remembered that there have been substantial investments in energy efficiency projects in the public sector under the utilities' EEP programs regulated by ANEEL. So far, these have been undertaken almost entirely on a 100 percent grant basis. This approach, while lovely for the lucky beneficiary, severely limits the volume of possible projects. Beyond demonstrating economic viability, these projects have done little so far to help transform the market for EE services in the public sector.

c.   Distortions in Pricing Which Adversely Affect the Viability of Projects

Low energy prices have not been considered to be a big impediment to making projects viable. In fact, distortions in energy pricing are quite limited in the market segments of interest. In these, fuel and electricity prices roughly reflect the cost of supply, with one curious exception.

There is a huge difference in the cost of peak versus off-peak electric power. Medium tension business consumers typically pay six to nine times more per peak period MWh (assuming the same load factor for contracted demand in kW). The basic consequence of this distortion is that the returns are increased on investments, which simply reduce contracted peak demand compared to those which increase efficiency of use. The starkest impact is on cogeneration projects relative to peak shaving gensets.

The main price distortions effecting the viability of energy efficiency investments appear in the ESCO's own "supply chain" rather than in the prices of energy. For example, there are high taxes and surcharges, especially in relation to formally employed personnel, as well as potentially high liabilities when discharging employees. In the retrofit and building maintenance businesses informality is rife because it substantially reduces costs. This makes it hard for a firm, such as an ESCO—which must work "formally" in order to have access to debt and risk capital financing to compete profitably.

d. *Credibility of ESCOs and of the Performance Contract Mechanism*

Credibility of ESCOs is regarded as the least important problem by the industry; however, the broader issue of the credibility of the performance contract mechanism may be more of a restraint than survey results suggest.

For example, the GEF approved a project in mid-2007, which seeks to accelerate EE building retrofits in Brazil, had an emphasis on chiller systems. The project, entitled "Market Transformation for Energy Efficiency in Buildings," has not yet been effectively launched but has ambitious targets. In terms of resources, by far the largest component is to provide capital for a fund to insure ESCO project performance in buildings.

Another example is the measurement and verification of the savings in energy efficiency projects. Editions of the IPMVP have been available in Portuguese since 1998. The president of ABESCO estimates that more than 3000 savings reports have been prepared by ESCOs over the years. Nevertheless, there has been an ad hoc approach to verifying results. Understanding of the protocol is not widely disseminated and very little is known, in fact, about the characteristics of all those savings reports. Lack of understanding and accepted standards for M&V could be more of a barrier in the future.

e. *Financing of Projects*

ESCOs site access to financing is the single biggest item inhibiting successful closure of project negotiations with potential clients. This confirms a widely held viewpoint that financing is the single greatest bottleneck for growth.

Most attention has been given to diagnosing and overcoming barriers to commercial bank loan financing to ESCOs and ESCO projects. Real interest rates for loans have been among the highest (if not the highest) in the world. However, the high cost of money has not been the main barrier.

A more important barrier has been the guarantees required for bank loans—especially those made with commercial banks intermediating medium term loans from the BNDES. Loans originating in the BNDES are the cheapest and often the only feasible source of medium term debt capital available to most firms. However, requirements for real "collateral" are very high while there is no effective value ascribed to a project's future receivables. The time needed for approval of loans has also been long. All this has made it impossible for most ESCOs to take out any loan at all. Meanwhile, consumers—faced with their own restricted access to credit usually prefer not to take out loans for this purpose. Loans to do building retrofits were unheard of until a couple of years ago and still are very scarce.

In an attempt to resolve this impasse, in 2006 the BNDES introduced an innovative new credit window for ESCOs called PROESCO. As noted above, PROESCO was designed to adjust credit guarantee requirements to better meet the needs of economically viable energy efficiency projects (see the box above). The program has been slow to start up and it is still too early to say whether it will meet widespread expectations that it will substantially expand access to debt financing for sound projects.

ESCOs certainly feel restricted by lack of risk capital. ESCOs indicated the most interesting innovation in financing would be the availability of private equity. It trumped the more predictable option of a mechanism to provide credit guarantees for performance contracts.

Unfortunately, there has been little advance in bringing private equity to the ESCO industry since a pioneering operation with one ESCO a few years ago. There are several difficulties which inhibit a larger entry of private equity including:

- ESCOs are seen as services companies. In general, fund managers, above all those of VC funds (in opposition to PE funds, which invest in medium and large-sized companies), are predisposed against investments in services companies. There is a widespread perception that services companies tend to grow at a slower rate than companies which commercialize products (whether tangible or not).

- There is also a perception that services companies suffer greater pressure on their profit margins, and that gains of scale are not easily practicable. In Brazil, there are specific problems with high taxes and rigid labor laws which create additional pressure on profit margins.

- The great majority of ESCOs are too small to be of interest to VC/PE funds. Due diligence costs are substantial and in order to dilute them, the investment needs to be above a minimum size.

- VC investors generally try to choose companies positioned to be leaders in their sector of operation. In this context, no fund manager seeks to invest in two or more companies which develop the same activity (for example, two ESCOs). Funds also seek to diversify their portfolio—thus it is very unlikely that any fund would develop a specific expertise in the EE sector.

- Funds need a plausible exit strategy for disinvestment, which probably means a strategic buyer; i.e., a larger company. At present there are few, if any, prospective buyers on the horizon, whether domestic or foreign.

- Most ESCO owners are reluctant to lose control and the due diligence process can be very off-putting. The ESCOs would usually prefer the option of SPCs (Special Purpose Companies) for individual projects. However, SPCs appear to be of limited interest for most EE projects, principally because they are usually too small to justify this approach.

- Finally, the difficulty in raising debt financing for projects has been the greatest impediment to the growth of VC investment in ESCOs and EE projects in Brazil. The cost of capital of VC funds is close to the average return of efficiency projects. Without leveraging the invested capital via debt, in a manner to reduce the average cost of the capital employed, equity investors are reluctant to enter the sector.

Another approach to bring private equity which was considered was the creation of a "trade receivables fund." Available in the Brazilian market since late 2001, trade receivables funds (in Portuguese, "Fundo de Direitos Creditórios," or "FIDCs") had grown steadily. Most FIDCs are issued for the public sector and/or large companies, such as electric utilities.

While there has been little entry of VC/PE, one does see more

ESCOs, which are subsidiaries of much larger companies. The entry of larger companies is another route to increasing private risk capital in the ESCO business.

Looking to the Future

The outlook for growth of the ESCO industry is positive and most ESCOs appear to be optimistic. There are both general reasons and more specific sectorial motives for this optimism.

At a general level, it is likely that relative macroeconomic stability in Brazil will be maintained, with relatively low inflation. In recent years credit has expanded rapidly. It will have to grow more slowly in the future. But measures to curb credit for consumption will probably not have much impact on the terms for ESCOs—which will be dominated by more specific issues discussed below.

Barring a collapse in a broad range of commodity prices Brazil's economy should continue to grow at 4+ percent/yr. Contrary to the situation of a decade ago, Brazil is well placed to absorb an external financial shock. At the same time, diverse infrastructure bottlenecks may inhibit growth much above that.

All this sounds "boringly normal" and that is the point. Such a sense of stability is a novelty in Brazil after decades of high inflation interrupted by numerous and disruptive anti-inflationary "packages." It provides a context, together with increasing competition, for more strategic planning by businesses and more attention to costs. This is conducive for ESCOs to sell their services and performance contracting more easily as time passes.

Other broadly favourable trends concern energy supply and demand and the environment. Energy prices are expected to remain high by historical standards, both for electricity and petroleum/gas, though they may not change much relative to inflation. Some petroleum derivative prices have been controlled and have not so far accompanied international increases.

In the case of electricity, there has been concern about the supply/demand balance three to five years into the future. There are unlikely to be actual shortages unless the economy grows faster than expected, but medium term prices could go up. This is fortunate, because from the perspective of energy efficiency in general, and ESCOs in particular, a fear of future rationing is a serious deterrent to consumers undertaking investments in most kinds of energy efficiency (except CHP). Why cut fat before the authorities set the baselines for rationing targets? Governments do not know (or do not want to know) how to discriminate between consumers who have made investments to improve their efficiency and those who continued to waste energy.

In the case of natural gas, the reliability of supply and of the price has been a problem in recent years, which has inhibited smaller scale CHP and distributed generation. This risk is likely to continue for several more years until planned new domestic production can come on line.

The potential for economically competitive EE retrofit projects is clearly large, though only roughly defined. This is the main driver for interest on the part of business consumers—ESCOs' main market.

At the same time, it is generally recognized that properly implemented energy efficiency measures present substantial environmental benefits compared to the equivalent expansion of energy supply. There seems to be a growing interest in, and priority for, environmental matters in the population of the country, especially among the middle class. Many larger business and financial institutions are concerned about their public image with regard to the environment.

Interestingly, the environment is the main motive for the involvement of financial institutions, including banks. When commercial banks address EE, it is usually through personnel of the environmental department. The same is true at the BNDES. The PROESCO credit window was developed by the Environmental Department.

The link of EE with environmental benefits is also clearly a major motive for greater interest in government at all levels in energy efficiency policy. It is probably the best lever for mobilizing political support.

Government support could make a substantial difference in the expansion of the market. A favourable outlook for several lines of official action is a big reason there is optimism among ESCOs about prospects. Among the actions involving the government, which could strengthen the development of the ESCO market over the next several years, are:

1. Opening the market for EE projects in the public sector—especially buildings and water & sanitation systems;

2. Exploiting the potential for greater impact of the utility wire-charge EEP program which was opened by the new guidelines published by ANEEL, in 2008;

3. The spread of PROESCO to increase access to debt from commercial banks for project financing; and

4. Deployment of the GEF/UNDP project "Market Transformation for Energy Efficiency in Buildings" being implemented via the Ministry of the Environment.

All these actions are feasible and some have begun. In principle, they should help to largely resolve many or even most of the barriers cited in the previous section, providing scope for accelerated growth from current conditions.

In each of these lines of action, however, it is possible that development of the possibilities will be slow and/or timid. As already observed, the start-up of the GEF/UNDP project to transform the market for EE retrofits in buildings is stalled. Also, the PROESCO credit line has been slow to obtain adherence from banks and to build a pipeline of projects. It is too early to judge, but there are reasonable doubts as to whether it will achieve the objective of overcoming traditional guarantee restrictions. It may yet make sense to create a specialized Credit Guarantee Facility along the lines proposed in Lima et al., 2005. The design of a guarantee facility would benefit greatly from the experience with PROESCO.

Perhaps the biggest near-term political opportunity, and risk, for Brazilian ESCOs resides in the development of the utility wire-charge EEP program overseen by ANEEL. Overall, Brazilian ESCOs are quite dependent on this mandatory program for revenue (if not performance contracts).

At a minimum, there is considerable inertia to leave things as they are for awhile, perhaps to see what utilities propose. The new rules require considerable adaptation even though they simplify the bureaucracy. Beyond that, it is not necessarily in utilities' interest to maximize the impact of the EEP or to lose control of important segments—as things are, all projects must be proposed by them.

There should be a sense of urgency about better use of the wire-charge resources because the basic enabling legislation comes up for renewal in 2010. It cannot be assumed that energy efficiency will continue thereafter to receive resources. During the last renewal in 2005, energy efficiency was almost eliminated.

The wire-charge for energy efficiency (and R&D) is only one of many surcharges ("encargos" in Portuguese) which, besides taxes, separate consumers' payments in their bills from the utilities' revenue. The cumulative impact of these surcharges is very large and has expanded in recent years. There is a general pressure from business associations representing consumers, and often from utilities, to reduce the total of these surcharges. In order to justify renewal, it would be helpful to be able to show programs with significantly greater impacts than in the past. Meanwhile, it also makes sense to use the resources while they exist to move towards transforming the market. It is very probable that the share of wire-charge resources going to EE will diminish substantially after 2010.

Fortunately, these factors, and possible government actions, which can promote the ESCO market, are diverse and setbacks with one or another may not impede advances. Nor does the growth of the ESCO business depend solely on initiatives involving the government. The prospect is for some improvement in the terms of doing business regardless of whatever happens in the short term. If the potential of this favourable moment is developed intelligently, growth could be far more vigorous..

Home-grown experience has been accumulated about diverse aspects of performance contracting, which so far has remained dispersed and little analyzed. There have been numerous performance contracts undertaken in Brazil in recent years, while there have been hundreds, if not thousands, of M&V plans implemented. To this author's knowledge there has never been a systematic review of this rich experience. Such a review would help to establish more exactly what business practices have, in fact, been effective. This would be a useful reference point for developing new approaches.

With diverse economic trends underway, promising government initiatives and possibilities to create new instruments to transform the market, the wind is favourable for the ESCO industry in Brazil. However, in order to reach the objective of a business which is close to achieving its economic potential, some clever navigating will be needed both by individual firms and by groups of stakeholders.

Chapter 9
Down Under

While the authors of this book suffer the winter blahs north of the equator, there is a wonderful place under the sun where Kiwis and Aussies live. When these folks look up they see a different sky, with the Southern Cross and other interesting configurations. They have koala bears, kangaroos, and "strange" flora and fauna. It is, indeed, a different place.

But it is also much the same. They worry about energy prices and wonder how to conserve energy and cut utility bills.

In this land of summer in December with 4 million Kiwis and over 20 million Aussies, energy programs are still needed. New Zealand imports 26 percent of its energy, consuming 4.6 m TOE more than it produces each year. By contrast, Australia's output is more than double its consumption and it is a net exporter.

Ironically, the energy efficiency (EE) activity and ESCO development do not mirror this need. New Zealand shows very little EE industry development while Australia has been relatively active for a long time.

NEW ZEALAND

In 1998, the New Zealand government took strong steps to promote the development of an ESCO industry. On behalf of the government, Hansen Associates conducted performance contracting seminars in Auckland, Wellington and Christchurch. The utilities were involved in this training and were looking to establishing their own ESCOs. Honeywell had recently executed a performance contract in a hospital in Tasmania and generously shared a detailed case study. An embryonic industry was forming.

Then, an election was held and the political leadership, including the energy agency, changed. Energy efficiency and the support for performance contracting did not appear to be a priority for the new administration. The efforts that were underway dwindled and do not appear to have rebounded.

Mr. Rob Bishop with Energy Solutions summarizes the present situation as follows:

> There's really virtually no ESCO activity happening in NZ at present. There have been attempts since 1994, but all the companies offering this have failed, due to lack of sales.
>
> This is at least partly due to the market being "poisoned" by an Australian company ... offering shared savings in the early 1990s, when the electricity market was being restructured, who got all the savings from changing the user to a better tariff, then taking thousands of dollars of savings (typically 50 percent) for the next several years for a few minutes of work.
>
> So people think shared savings are "too good to be true," and won't go along. People have a long memory about that type of scam. I've tried for years to sell projects on the savings with no success.
>
> Now I guarantee the savings from my projects (at a three year "payback" or better, or else I refund my fees to make up the difference), but I know I can always achieve a one-year or better payback (doing HVAC operational improvement, and charging low fees), so that's a good safety margin.
>
> Apparently Honeywell has sold some projects here, but based on covering deferred maintenance. I know of one project with Department of Courts, and one at Otago Hospital. There may be others.
>
> In my opinion there is presently no indigenous ESCO activity in New Zealand. Perhaps this is part of why there is so little improvement in energy efficiency/productivity (currently about 0.4 percent per year, to my understanding), driving the present surge in energy generation and transmission construction.

AUSTRALIA

Dr. Paul Bannister has provided valuable information on ESCO development in Australia and assessed its current status as presented below.

Before discussing the ESCO industry, a brief look at its renewable efforts seems warranted. Australia, along with the US and Canada, was the recipient of a lot of criticism during the Kyoto Protocol negotiations and was portrayed as hostile to reducing greenhouse gas emissions and energy conservation in general. Australia has actually made renewable energy an essential part of its low emissions energy mix and has had a small but growing renewable energy industry. The renewable energy base has predominantly delivered hydroelectric power into the national grid. It also has had the availability of biomass which enabled a range of small scale generation projects to be developed. These activities provided the springboard for new projects, which included efficiency gains and the refurbishment of the existing infrastructure – made viable by government incentives.

With a relatively small population, the commercialization of some technologies is not as economically viable. Without a large market to foster production of large quantities, Australia is, too often, not competitive in new technology pricing. Despite these drawbacks, there have been some renewable energy manufacturing successes, especially in solar energy.

Development of the ESCO Industry

Energy performance contracting (EPC) in Australia began in the late 1990s. The initial activity in the sector was strongly driven by the New South Wales (NSW) state government, which sought EPC for many of its hospitals and some of its other facilities. A number of universities were also involved in the procurement of EPC in this period, as well as a few smaller contracts in other states.

Early contract processes were characterized by relatively long and drawn out contract negotiations and a high percentage of sites did not proceed to contract. This was probably due to the inexperience of the participating ESCOs and the clients; the ESCOs in terms of not necessarily inspiring the appropriate level of confidence and the clients in terms of the selection of suitable sites or a lack of understanding as to the process. Another problem was the lack of consistent methodologies by ESCOs, with different contracts, guarantees, etc., so customers found it impossible to compare offers. Nonetheless a small number of significant EPCs were commenced, focused mainly on hospitals in NSW, but also including commercial and industrial facilities.

In parallel with this process, the industry formed the Australasian Energy Performance Contracting Association (AEPCA, www.aepca.asn.au), to develop a consistent procurement methodology, and undertake industry development, including marketing. AEPCA was able to attract federal government funding to enable the production of best practice guides. The association also developed a model contract to address the difficulties encountered in the initial period of activity. These tools, especially the best practice guides, have attracted interest from a number of international bodies.

The major barrier to growth of the EPC market, however, was and continues to be, a lack of consistent government programs and policies encouraging energy efficiency. As discussed, NSW took the lead initially but, with changes in NSW government focus after 2003, activity in that state largely ceased. Queensland has initiated an active program which is on-going; however, there has been virtually no activity by any other state or the federal government. Fortunately, in 2007-8, activity has increased on a number of fronts as discussed below.

Current Activity

There are signs of a promising future for the Australian EPC industry due to a number of factors, including:

- The Clinton Climate Initiative, which is leading to activity in the state of Victoria in particular;

- A surge in interest in the use of cogeneration and trigeneration as a means of lowering greenhouse gas emissions (relative to Australia's generally very intensive greenhouse gas grid supply at 1-1.5 kgCO_2 per kWh);

- The recent signing by the new federal government of the Kyoto Protocol and new policy development as a result; and

- Private sector initiatives for the provision of energy services, mainly in the commercial sector.

Nonetheless, the EPC market remains relatively under-developed in Australia.

Market Participants

There are a number of ESCOs currently operating in Australia. These are dominated by equipment suppliers and control companies with some participation from utilities and independent ESCOs. Key players include Honeywell, AGL, Energy Conservation Systems, Total Energy Systems and Dalkia. Other control companies such as Johnson Controls and Siemens are also targeting new work in this area.

Contract Types

The majority of contracts are based on cost savings rather than energy efficiency. They do offer savings guarantees. The domination of the public sector in the client base has meant that the majority of the funding has been client-provided rather than ESCO-provided.

A new field of activity in the past year has been the provision of cogeneration and trigeneration plants on a chauffage basis. This has achieved some uptake in the commercial sector where local initiatives are placing a strong emphasis on the achievement of large reductions in greenhouse gas emissions. The combination of this technology, its relative complexity of operation and the high costs involved favour ESCO-funded chauffage contracts.

Market Sectors

For conventional EPC contracts; i.e., where the primary technologies look for reductions in energy and water use, the majority of ESCO activity has occurred in the commercial and institutional sector. To a large extent, the market continues to be dominated by state-government procurement. Private sector and federal government activity is very limited. A notable exception is that one ESCO has teamed with a major commercial portfolio to provide performance guaranteed upgrades to tenant lighting. As a result the technical focus of projects has been building services, such as air-conditioning, lighting and cogeneration/trigeneration.

Activity in the industrial sector continues to be dominated by one ESCO, which rolls up the 'conventional' EPC product into more comprehensive asset management type contracts.

A significant related industry is also appearing somewhat independently of the traditional ESCO/EPC process. It has become relatively commonplace for new construction projects and major refurbishments to specify a performance target using the Australian Building Greenhouse Rating scheme (now known as NABERS Energy (see www.nabers.com.au). This effectively serves an absolute operational performance target. This responsibility is generally being passed on to builders, who are unwittingly acting as ESCOs in this context. In some cases, such as government procurement of facilities on a lease-back basis, the building owner is taking on a 10-15 year obligation to maintain the same level of energy/greenhouse performance. This type of activity is increasing rapidly.

Barriers

There are a number of barriers to the EPC industry in Australia:

- **Government policies**: At the state and federal level, significant reductions in energy use or greenhouse gas emissions are not yet required. Recent developments relating to climate change should see this improve over the next two years.
- **Government procurement process**: The accounting rules for state and federal government largely prohibit the use of private sector financing. This makes it hard for ESCOs to offer off-balance-sheet solutions to government bodies. Furthermore, normal procurement guidelines make the process of selecting an EPC contractor laborious and expensive for the contractor, with outcomes often determined on relatively poorly justified grounds (such as the level of savings offered based on a walk-though audit). This is discouraging many ESCOs from participating in this part of the market.
- **Energy costs**: Energy costs have been traditionally very low in Australia. On the eastern seaboard, prices in the region of USD 0.06-0.09 per kWh are common for medium to large sites, which increases payback periods and reduces project economic viability. Gas contracts are also not structured in a manner that encourages energy savings, as charges are structured around pre-committed purchase levels that need to be renegotiated before significant savings can be realised.
- **Contractual issues**: In spite of the presence of an industry-backed model contract, the process of going to contract for an EPC continues to be very slow and potentially expensive. This increases the upfront costs for ESCOs and also has led to a number of contracts never being signed.
- **Maintenance**: Unlike in some countries, EPCs in Australia are not normally integrated into total asset management services, but are instead focused on one-off energy savings projects. This means that the site maintenance may be conducted by the client. This can lead to difficulties caused by changes in site operation and maintenance affecting performance outcomes, particularly as the project ages. Furthermore, with the resultant lack of daily contact between ESCO and client, it can be difficult to maintain long term interest in the contract, resulting in a breakdown of contractual cash flows, communication and results.
- **Lack of human resources**: Australia has a chronic shortage of skilled people working in energy efficiency generally. This has made the

expansion of a skills-intensive ESCO industry difficult, particularly when there is currently good business in more traditional service areas.

Industry Drivers

There are several factors in operation that are favourable to the EPC/ESCO industry:
- *Grass-roots level interest in global warming*: While the Australian federal government has scarcely been a leader in international activity on climate change, this very lack of leadership has spawned a significant level of activity in the state government and private sector directed towards the reduction of greenhouse gas emissions. While little of this currently takes place under an EPC arrangement, there is significant potential for this to become an area of increased activity;
- *Emissions trading*: The Australian federal government is developing an emissions trading scheme which is intended to be implemented in 2010. This should be expected to increase general activity in the energy efficiency sector and should also prompt increased interest in guaranteed savings;
- *The Clinton Climate Initiative*: Melbourne is a participant in this initiative and there is already evidence that this is increasing ESCO activity in this region; and
- *Industry documentation*: AEPCA has developed best practice guidelines for the procurement of EPCs and for measurement and verification that assist the uptake of EPCs in the market.

Outlook

In the next three to five years, it should be expected that the EPC industry will experience primarily organic growth. As government support for the EPC industry is not particularly active, most ESCOs appear to be privately promoting EPCs as one of a number of service models within their business, rather than as a sole means of trading.

Nonetheless, the rapidly expanding activity in energy efficiency, the introduction of emissions trading, increased interest in cogeneration/trigeneration solutions and specific activities, such as the Clinton Climate Initiative, will mean that the industry, as a whole, should be expected to grow significantly.

Chapter 10
The Global Picture

*I*n your hands is the dedicated work and valuable insights of many, many people from around the world, people who share an interest and belief in energy performance contracting. Thanks to them, we are able to offer the first global assessment of performance contracting and the energy service industry. The  richness of their stories, the obstacles overcome and the lessons learned offer a compendium of exceedingly valuable guidelines for those striving to create or strengthen our ESCO industry.

As advocates for performance contracting around the world, gathering information for this book has been a very gratifying experience. There certainly is a way to go, but the maturity of the ESCO industry in so many corners of our world is a joy to behold. ESCOs are alive, well and growing!

The unique contributions performance contracting makes to improved energy efficiency, economic growth, and the conservation of our natural resources is widely recognized. Documentation is growing that energy efficiency (EE) is the most cost-effective way to reduce environmental pollution. Performance contracting has been recognized as a valuable tool in our efforts to control global warming and to achieve a healthier environment. In fact, it is becoming evident that energy efficiency is an extremely desirable way to finance our growing desire for more renewable energy. Further, ESCOs make a natural marketing channel to combine both worlds.

While each story is unique, from the experiences of each country come some common threads. This chapter considers those threads as they are reflected in the problems, needs, barriers, and opportunities that prevail throughout the world.

COMMON THREADS

Woven throughout the reports from our colleagues in all corners of the globe are similar concerns. For example, there are repeated statements of the need for quality measurement and verification procedures. The lack of awareness within the potential client base of what ESCOs can do is prevalent and critical. The most dominant theme, however, is the role that governments play, which clearly has, and will continue to encourage, and/or hinder industry growth.

Government Role

Governmental actions, such as those depicted in the report from Japan, reveal the vital role governments can play in fostering energy efficiency and the development of energy service companies. Other country reports indicate that governments, which fail to act, do not foresee unintended consequences, or do not appreciate the opportunities incentives offer, can inhibit ESCO industry growth.

Governments have even opposed performance contracting, which was the case in the early days of ESCO development in the US. Fortunately, strong ESCOs working with state governments were able to overcome this initial barrier. Today, of course, the US government is an active proponent of performance contracting. The US has not been alone in this struggle, a province in China and the Treasury in the UK have declared at various times that energy performance contracting was illegal.

Part of our "illegal" difficulty has been our legal profession. The contracts required for ESCO projects do not always fit within the lawyers' experience. In many countries, the lawyers have been reluctant to embrace a contract that is unfamiliar. Judicial systems which do not work, or work very slowly, also create legal problems. Laws and regulations often work against performance contracting and getting changes made can be a very slow process. There is much to be said for the "informal understandings," more prevalent in developing countries, which the report from Thailand describes. The premise is that it is who is involved that is important; not the words on paper.

A surprising number of countries are waiting for the government to take some action. The lack of references to ESCOs in the Polish legal system leads to "enormous reluctance and apprehension among potential customers." In Vietnam, the potential industry, as well as its customers, are apparently waiting for the government to establish guidelines and directives. On the other side of the world, our Greek contributor observes, "The lack of ESCO business is based on the absence of a positive and institutional environment for the initiation and viability of an ESCO operation." To compound the problem, Greek laws until recently prohibited a private body from operating or managing a building's energy service infrastructure for government owned facilities. And from northern Africa, experience suggests that government action to build reliable and sustainable markets might be required to save the ESCO industry.

The tendency to rely on government to act, however, can be pernicious. In a country such as New Zealand, where 26 percent of the energy is imported and there is no "indigenous ESCO activity," one wonders why the government has not taken the lead.

Some governments intending to do good, do harm when it comes to supporting the development of an ESCO industry. Brazil offers an excellent example. In its desire to reduce energy consumption, the government provides 100 percent grants for energy efficiency (EE). With the best of intentions, tax dollars are being used to undermine the private sector's attempts to serve this need.

Some governments do good without a conscious effort to do so. Moving from a history of subsidized energy prices, especially in command and control economies, governments have changed to actually taxing energy. While such taxes might hurt the international competitiveness of the general economy, they do increase energy prices, making EE efforts more attractive and serving as an incentive to ESCO growth. Other governments inadvertently support ESCO development. A classic example is the US government's tendency to issue mandates without accompanying capital to meet new government requirements.

Italy has developed a different incentive program, which gives registered ESCOs EE certificates (white certificates). The monetary value of these certifications has been credited with the increase in the ESCO market.

Changes in culture or the national context in which an ESCO industry wishes to operate may create barriers or more attractive operating conditions. Historically, all the Chinese sectors related to EE projects were implemented by administrative measures, from organization, technology and investment. Today, a new context for ESCOs exist, "… with the withdrawal of the planned economy and the development of a market economy, the investment and operation behavior of enterprises are becoming centered on economic benefits. The past administrative measures do not work effectively any longer. Therefore, building up the market-based mechanisms of energy conservation and efficiency, the energy conservation service industry has become the most urgent need of the nation."

When government roles are mentioned, the tendency is to think of federal actions; however, local governments often play key roles. The growth of the ESCO industry in Germany (the most active ESCO market in Western Europe) credits "local political support and individual drivers" for its rapid growth.

Government actions clearly influence ESCO development around the world. It is satisfying to end this consideration of the governmental role by looking at the government efforts in Japan, which have repeatedly proved key to ESCO development. The authors of the Japanese report offer a clear scenario of the step-by-step actions the government has taken to move toward its EE and environmental goals. In the home of the Kyoto Protocol, it is gratifying to note that several of those federal efforts recognized the crucial role ESCOs can play in achieving the environment goals envisioned in the protocol.

Accreditation and Certification

Closely aligned with the government role in ESCO development is the field of accreditation. Not all certification procedures, however, are government based. In the US, the National Association of Energy Service Companies has a certification procedure for its member companies. This grew out of a pioneering accreditation effort by the Canadian ESCO association.

If we lump a variety of accreditation and certification activities together, a pattern emerges of attempts to assure a level of quality and the fostering of confidence.

- In Austria, the development of certification of ESCOs and ESCO businesses is credited with the industry's growth. The Graz Energy Agency issues a "Thermoprofit" quality label, which offers a guarantee of reliable high quality ESCO proposals. In addition, its "ecolabel" denotes quality of ESCO services and compliance with Thermoprofit standards.

- The National EE Agency of South Africa has a broad mandate to promote EE throughout the country. Among its responsibilities is the accreditation and development of the ESCO industry.

- A new law in Turkey provides a framework for ESCOs and defines the qualities for accrediting ESCOs as well as certifying energy managers and auditors.

- Under a donor/government program to promote ESCOs in the Philippines, the first recommendation is to "Develop and institutionalize an ESCO accreditation scheme, …defining and categorizing different types of ESCOs."

- Energy savings measures implemented by ESCOs in Italy must be certified by the Market Operator, which issues certificates at the request of a government regulator. These "White Certificates" acquired by ESCOs can be sold to distributors.

The above excerpts from various reports suggest a growing inclination to use accreditation and certification in the industry. There is an obvious need for some quality assurance; however, it is not always clear whether these efforts are designed to reassure customers or governments. At this point, it is also unclear how the effectiveness of these accreditation and certification procedures are measured or policed.

Measurement and Verification

The development of some means to measure and verify energy savings in most countries to date has been done with government involvement.

The contributing authors frequently mention the need for some type of accepted means of quantifying energy savings. Difficulties in getting project financing in Tunisia, for example, are reported to be directly related to "unfamiliarity with assessing energy efficiency investments…" and the "lack of information on the benefits of energy efficiency measures."

In discussing barriers to ESCO development in Malaysia, it is noted that the need to document success exists in many countries:

> Greater than all the concerns stated above is the reluctance of a company to invest in the energy saving project when the first project did not prove itself to be a good investment. After granting the first project and waiting until it was implemented, the management still did not have a chance to see any actual savings, as the in-house engineers were not equipped with the needed M&V skills to document savings.

The measurement and verification (M&V) protocol advanced by the US, the International Performance Measurement and Verification Protocol (IPMVP), was initially a government action. Responding to the need for some type of measurement of savings, various methodologies were emerging. Multiple versions were confusing and not necessarily accepted by another party. The US Department of Energy gathered key people together and a consensus evolved into what an M&V procedure should be. After two previous versions had been published, in 1997 the first IPMVP became the "international" protocol. Adoption of the protocol, sometimes with in-country modifications, has been actively solicited and the IPMVP has become broadly accepted around the world. (The organization, which issues the IPMVP, is the Efficiency Valuation Organization (EVO) and the IPMVP documents are now available at evo-world.org.)

Sometimes this protocol is modified to meet local needs, as the authors of the Japanese report stated, "Although IPMVP provided the fundamental framework and the technical contents of measurement and verification, it was revalidated and adjusted to the conditions peculiar to Japan."

One of the problems ESCOs face is the lack of historical data to provide a baseline. In a study conducted for the Indian Renewable Energy Development Agency Ltd. (IREDA) and funded by the World Bank, the report noted that M&V procedures are a major barrier and recommended,

> The design and implementation of a fair and easy to understand M&V protocol is required ... M&V is a contractual issue and it is usually not a complex one. ... The M&V relies also on the quality of the baseline preparation. The main barrier at that level is inadequate data rather than the complexity of the issue.

Sometimes the problem is rooted in formerly planned economies where the lack of data to construct baselines, subsidized energy prices and poor management/operations of buildings have been common. The Czech Republic report summarizes the problem, stating, "Detailed, reliable information on present energy consumption and condition of buildings is lacking. ..."

To have effective M&V there must be a reliable baseline—a mutually agreed upon reference point. Further, that baseline must be more than consumption data, it must include information on the conditions that caused that consumption. The problem is not necessarily a bookkeeping problem; but, as in the case of Hungary, many areas do not have detailed billing systems and pay average fees per month. ESCOs are left with spending one to two years accumulating the baseline data, or relying on different ways to estimate the savings.

In South Africa, electricity tariffs fund a program that requires an independent assessment and reporting process for M&V before the National Electricity Regulator approves the budget. The M&V functions are outsourced to South African universities.

Too often, ESCOs create their own M&V methodologies and, as in Mexico, the procedure is not universal; so it may not fit other projects and the data may not be accepted by others. In some instances, donor agencies have become involved in modifying the M&V process. In Egypt, the UNDP, as part of its involvement in an EE project, produced a simplified version of the international M&V protocol. As yet no records are available as to how well the simplified version is fulfilling the requisite M&V needs.

M&V accuracy is expensive. The IPMVP reportedly creates findings predicted to have ±10 percent accuracy. It becomes a matter of cost vs. accuracy. Efforts to become more accurate must be weighed against the cost, as too much detail and/or precision can quickly become a prohibitive cost burden for the project.

From Germany comes this observation, "A serious problem for ESCO projects is the need to measure and verify savings, which requires a relationship of trust between the ESCO and the client." When the economic viability of a project is based upon something as nebulous as future energy savings, the relevance of trust becomes a major factor. When governments visibly support, incentivize, even mandate ESCO activity, this "trust" is thrust upon the public. Nevertheless, many country reports reveal an inclination of the public to accept such decrees and to move toward displaying the needed trust.

Financing

Problems in securing the needed project financing are pernicious and persistent. Often the money is there, but it cannot be accessed for project financing by an ESCO.

Before the range of financial problems revealed in the reports is addressed, it is noteworthy that some countries do not have financing problems. ESCOs in Japan, for example, are frequently owned by large companies and the parent company provides the needed financing. Hungary is representative of a few countries where, "financing of ESCO projects by banks is not a problem." The contributing author goes on to state, "On the one hand, some (mainly multinational) ESCOs have sufficient financial means. On the other hand, third-party financing is a well accepted and widespread scheme. Banks are particularly open to participate in performance contracting." (It should be noted, however, that the reports from countries like Japan and Hungary preceded the worldwide financial crisis that struck in October 2008.)

More typically, however, our contributing authors have pointed to a number of financing difficulties. An over-arching difficulty is presented in the Italian report, "The lack of interest from financial institutions has been a significant barrier." This "lack of interest" is often paired with a lack of understanding of energy efficiency as reflected by a banker in Moscow, when he stated "We don't lend money for hot air."

The bureaucratic problems that create burdens for the financial institutions and/or the applicant also constitute serious barriers. Poland offers a representative example of the problems that can be caused by tax laws and accounting regulations. First, the owner must pay taxes upon issuing an invoice, regardless of when the client actually pays, or if he pays. Also the total financial value for the whole period of construction and financing services is reflected in the enormous VAT sums ESCOs must pay. In addition, for thermal modernization, the customer must make a first installment payment equal to 50-80 percent of the entire project.

Financing difficulties can also be traced to ineffective communications. Banks are unaware of the potential benefits of ESCO ventures and the potential borrowers (ESCOs and/or clients) fail to make the case. In Malaysia, the EE projects are perceived as high risk and low-return ventures and, therefore, are questioned as to whether they are bankable or feasible.

The assessment of current ESCO financial conditions in Mexico by Monica Perez reveals that Mexico offers a microcosm of financial conditions facing ESCOS around the world. Recognizing the international prevalence of financing problems, such as those in Mexico, the International Energy Efficiency Financing Protocol (IEEFP), as currently developed by EVO, was formed to provide commercial banks the technical expertise and procedures to establish cash flow financing as an accepted procedure. Perez has been responsible for IEEFP development work in Mexico, and has summarized this work:

> IEEFP-Mexico is a series of tools and best practices that will be used as a blue print to train local financial institutions (LFIs) on how to evaluate and assess the real risks and benefits of energy savings projects (ESPs). IEEFP-Mexico

is an effort led by Thomas K. Dreessen, chairman of the Efficiency Valuation Organization's (EVO) IEEFP Committee, and it was initially funded by the Asia Pacific Economic Cooperation (APEC) and is currently supported by the Global Opportunities Fund of the United Kingdom. The main goal of the IEEFP is for LFIs to have a better understanding of how the cash flow of ESCOs can become the main repayment source for ESCO loans. In the long run, this program is expected to result in a considerable increase in participation of LFIs in ESCO financing due to a reduced risk perception. Also, several other stakeholders will become more involved and interested in this market once there are standardized practices with decreased risk and transaction costs. The market will become sustainable and will not be in need of support programs.

Given the problems ESCOs have had in securing bank financing in so many countries, the progress of the IEEFP in Mexico is worth watching and, hopefully, replicating.

Financing Mechanisms

The shared savings and guaranteed savings models discussed in Chapter 1 are prevalent around the world. In Canada and the US, guaranteed savings is the predominantly accepted model. Outside North America, the shared savings model seems to be in greater use. Exceptions are South Africa and the Czech Republic, both having ESCOs that rely more heavily on guaranteed savings. Thailand's ESCOs, who have had problems with over extending credit with the shared savings procedures, have taken to using the special purpose entity model to reduce the leveraging.

There are obvious reasons a shared savings model is often preferred as an ESCO industry first emerges in a country. Initially, clients are more attracted to a new concept if they do not have to incur the debt obligation. Second, in many transitional economies the potential client has not established the creditworthiness required by banks to incur the credit risk themselves. Third, the large ESCOs have deeper pockets and can more easily handle the leveraging associated with shared savings. In fact, shared savings can keep the small ESCOs from gaining much of a market position.

Modifications to the more standard financial mechanisms exist. Four versions of interest are:

- Dalkia's undisclosed third-party financing. The ESCO has an informal agreement with a preferred FI, which steps in only in the event the client defaults.

- FEDESCO in Belgium follows the Econoler first out model. This approach is apparently attractive in Belgium as the ESCO-based financing limits participation to responsible partners for the entire project.

- In Cote d'Ivoire, the government provides each ESCO USD 10,000 in seed capital for equipment, living allowances and initial overhead. In addition, a government agency administers a revolving fund, which helps fund the end-user's investment and helps guarantee partially, or totally, a loan by a private commercial institution to an end-user under contract to an ESCO.

- In Kenya, the ESCO charges near market fees for project development as well as "hand holding" fees throughout the implementation process. In addition, an agreed upon "bonus" payment is paid at intervals based on the verified energy savings realized.

One of the financing issues that appears to remain unresolved in several countries is the owner's inability to keep the energy cost savings. When budgets are reduced by the amount of utility costs avoided, as it is in Mexican federal buildings, the occupants' incentive to reduce energy consumption is lost, nor is there any money to pay the ESCO for the services it has rendered. To meet this need some ESCOs have become "utilities" by redefinition, taking all the savings, paying the power companies and then using the excess savings to meet the owner's specified needs. In the US, the Department of Defense had a provision included in its annual national defense administrative act that stated all the savings would no longer all go to the general fund, but instead one-third would go to the base commander for base recreational purposes and one-third would go to the base O&M budgets. Not surprisingly, in the early days of performance contracting in US federal facilities, all but one contract went to military installations.

A happy spin off of this resolution was the inclination some US ESCOs have to include in the contract, a provision that provided a certain portion of the savings would go to O&M needs. Historically, O&M personnel frequently viewed performance contracting as a threat or intrusion and would actually flip those switches and turn those knobs so the ESCO project would not deliver as promised. Dedicating funds in the contract not only reduced O&M resistance to a performance contract, but actually encouraged the O&M personnel to make the project work.

These modifications are typical of the inventiveness displayed by emerging ESCO industries, when they find themselves needing to adapt to existing conditions in their respective countries.

Salted throughout the various commentaries is evidence that management views EE as an investment to be compared to investing in increased production. The problem is serious enough to prompt the following statement from Uruguay to be underscored *"Businesses in Uruguay, as elsewhere, tend to invest in capital equipment to increase output rather than reduce costs."*

From the US, the problem has been identified as follows:

> The industrial market is hindered by the policy of most American manufacturing companies, which results in project payback requirements of typically less than two years, which preclude comprehensive EPC projects employing multiple technologies. Top management frequently fails to understand that financing energy efficiency out of avoided utility costs is not the same as requiring new budget allocations.

In the list of Thai barriers, the concern emerges as, "The project return on investment is not considered attractive enough to proceed when compared to their conventional business capacity expansion or capability (product line extension). This impacts budgeting for both capital and resource requirements."

To overcome this kind of thinking, ESCOs around the world must learn to differentiate EE investments and increased production investments. ESCOs need to make the business case for EE. What most managers and investors don't seem to realize is that the financing source for the EE investments is right there in avoided utility costs. This is wasted money already in the budget. As long as inefficient measures are in place, that money is going up the smoke-stack. While money is being wasted, it is also creating more pollution every day.

In contrast, the money to increase production is new money that must be added to the budget. On one hand, the EE investment reduces operating costs. On the other, increased budget allocations for increased production must be passed along in product unit costs, weakening a firm's competitive position.

When managers around the world talk about internal rates of return, or they declare that the EE investment must meet production cost ROI parameters, it is clear they do not understand the distinction between these two sources of money. ESCOs need to understand and then convince management that by not acting on EE opportunities, managers are burning money as sure as if they put a match to it. Burned money is totally wasted; never to be recovered.

Banks

Understandably, banks typically are interested in low-risk clients with good potential for profit. Bank personnel tend to be traditional and follow prior practices. This means they have an asset-based mentality and focus on the creditworthiness of the client or ESCO. For example, bank loan requirements are very high for real collateral in Brazil and there is no value ascribed to a project's future receivables.

In some countries, these problems are compounded by other conditions. For instance, Egyptian banks show little enthusiasm for using energy measures as collateral because once installed the measures become part of the real property. A more prevalent problem is the reluctance of banks to loan money on predicted energy savings. Banks frequently lack the internal expertise to evaluate EE projects and the associated risks.

The problem in most countries around the world is not the lack of money, but an inclination of banks to institute credit conditions that cannot accommodate the financial arrangements of ESCO projects. The length of the loans and the perceived risks have caused interest rates in various Chilean banks to fluctuate between 8 and 33 percent.

The Indian experience sums it up well, "The availability of money is not as much an issue as the terms on which it is available."

Customer Awareness of ESCO Opportunities

Among the pages of this book can be heard the plaintive cries from ESCOs who hunger for clientele that have some idea of EE needs and what ESCOs do. ESCOs in the UK lament that even after 25 years, the potential client base does not really understand the process.

Ignorance takes on different characteristics. The "informed" public seems to range from total unawareness to those turned off by excessive claims of success, tariff correction "rip-offs," inadequate audits, weak M&V and project failures. When a procedure like performance contracting is not understood, bad news travels fast, problems are more readily believed, and there is an obvious reluctance to listen to explanations or positive information.

A company's faulty work may damage the market potential in a whole country. Repairing that damage takes time. Carefully orchestrated marketing strategies are often needed and information about successes must be widely circulated.

Out of a lack of information comes false expectations. It is not unusual for people to view an ESCO as a type of bank, set up to finance an existing audit. Too often ESCOs are expected to implement, even guarantee, another auditor's work. This creates tensions and misunderstandings; resulting in diminished trust. Potential customers should realize that an audit from an unknown source, which an ESCO is expected to implement, operate and guarantee savings, creates considerable risk to the ESCO. Typically, the ESCO will conduct another audit and the owner, whether he knows it or not, pays for two audits. This starts the trust issue festering and it can even be a deal breaker.

There are three major conduits to develop customer awareness. First, it can be established by government support, demonstrations and good dissemination strategies. Demonstration projects, which utilize typical measures with results tallied by a third party, can help a great deal. First efforts may be modest, but should always be well documented. Second, the collective voice of an ESCO association can help fill this void. The association must, however, see information dissemination as a major function if it is to fulfill this need. An association that only claims to inform leaves its members assuming this need is being met while the chasm broadens.

The third crucial area of developing awareness is a concerted marketing effort. Marketing must address the initial need for the potential client base to learn of the opportunity, but must continue to foster that awareness as the industry emerges. Using happy customers to inform their colleagues can be very effective.

Brazil and Thailand both underscore the critical role customer perception plays in furthering ESCO business. In spite of the fact that ESCOs stand ready to implement energy savings measures at no front end cost to an owner, and the owner will gain lower operating costs and new, more energy efficient equipment, the customers lack of expertise and their risk perceptions can constitute a major obstacle.

For an ESCO from outside a given country wanting to enter the business, understanding the cultural expectations is critical. For example, how negotiations are handled can be key. ESCOs entering the market may find there are rituals, even an art form to negotiations that requires patience. Comments about "carpet salesmen" negotiators in North Africa, provides exactly the insight an outside ESCO needs to understand and communicate with customers. Understanding whether the negotiations take place within the context of assumed distrust as in Europe, or assumed trust as in the US, can make a huge difference.

Associations

In addition to its effective dissemination practices mentioned above, an effective ESCO association can, and should, serve a number of valuable purposes. The association can become a forum for exchanging best practices and resolving problems. Its collective voice can also make industry concerns known to the government—and can inform its membership of new government actions. Germany's work with the judicial system serves as a good example.

An ESCO association can serve a valuable function, training and (at times) re-training the new ESCOs. Too many emerging ESCOs-dubbed-WISHCOs in the industry do not do their homework and their initial faulty efforts can damage the credibility of the entire industry in a country.

The review of activities performed by the Japanese Association of Energy Service Companies presented in Chapter 6 offers a comprehensive, detailed list of potential association functions. The growing number of ESCO associations around the world offers opportunities for mutual benefit. To this point the Asian associations have met to examine ways they can establish a working network.

Renewables and Alternative Fuels

While there are passing references to renewables in the reports, there are few substantive discussions of ESCO activity in this area. An examination of the reports suggests four reasons.

1) Renewable energy is often expensive and frequently does not have a place in the economically viable package that ESCOs struggle to build;
2) Renewable proponents often place their advocacy in competition with energy efficiency; so they are not always viewed as compatible activities;
3) Renewables are often a mass effort, such as wind farms, and frequently are not viewed as a customer-specific service an ESCO might offer; and
4) ESCOs may not have the expertise needed to incorporate renewables into their offerings.

EE has played a key role in meeting Brazil's environmental concerns and Mexican ESCOs have sometimes taken on the role of carbon market brokers, but "green" ESCOS are scarce.

A statement from the US, "EE Pays for Green" offers the best reason ESCOs should explore this opportunity. Energy savings can help buy down the more costly renewables. ESCOs can provide a valuable service to those striving to be more green by marrying the two worlds; putting EEs and renewable energy in the same package.

CONCLUSIONS

Pierre Langlois, drawing on the extensive global experience of his firm, Econoler International, was asked to assist in framing some joint conclusions regarding this work and the future of energy performance contracting (EPC). Reports from 49 countries gave us much to think about and pointed to a very promising future.

EPC and the concept's delivery agent, ESCOs, have come a long way since they were introduced to the market more than 25 years ago in the UK, the US and Canada. Even though it has been a known concept in some ways in European countries, the concept really got started in the form that we know today by innovators, who saw an opportunity in selling a result instead of just a service or a product.

Since that time, the concept has become more refined and the legal and financial mechanisms more complex. The attractiveness of the concept has remained and grown during these 25 years, even though there has been some ups and downs over that time. EPC is now known and used across the continents, in many different countries, and in different forms adapted to varying legal, cultural and financial realities.

Through our work, we have had the good fortune to exchange ESCO information with many of the prime actors of the EPC market around the world. We have learned much on the current state of EPC and on the ESCO activities in 49 countries. We are convinced that some other innovative ways to use the EPC approach are certainly in place in some other countries or regions that we do not suspect at this time, but our review will help the readers understand how it has evolved to date.

Probably the greatest finding in assembling and assessing these reports was the discovery of the spread and depth of the market and the many different ways EPC is used in all those countries. Many countries around the world are showing new ways to the world in using and adapting EPC. We can say with confidence that EPC has become a clearly established industry with a far reach.

While it will continue to grow and evolve, EPC will certainly change in character, but the basic elements of special services and guaranteed results will remain its hallmark. We can expect that carbon financing will start to play a role in the use of EPC, as ESCOs will see this new market as a great opportunity to improve on their current offering. We can imagine that the increase in energy costs over time will invite many end users to inquire about the concept and find it attractive enough to request proposals from ESCOs. We also believe that governments, while facing so many challenges related to energy security, the environment, and budgets, will see the EPC advantage and will make the necessary legal modifications to enable all public sector entities to use it efficiently. And we fervently hope financial institutions will become more comfortable with cash flow financing, thus providing key support for local ESCO industry growth.

Finally, we can only hope that this book will encourage the different stakeholders around the world to continue to promote and to develop energy performance contracting as one of the best mechanisms to help address the future energy and environmental challenges that we, as societies and private sector entities, face in the 21st century.

References

An important reference for this book has been the *Latest Development of Energy Service Companies across Europe* authored by Paolo Bertoldi, Benigna Boza-Kiss and Silvia Rezessy. This reference is published by the Institute for Environment and Sustainability of the Joint Research Center of the European Commission; reference EUR 22927 EN – 2007. The references cited below for Chapters 2 and 3, Western and Eastern Europe, are taken from this source.

Many of our contributing authors also cited references upon which their material was based. Those references are presented below in the order of the chapters and then in alpha order by country.

Chapter 1. ESCO Development
Hansen, Shirley J. 2006. *Performance Contracting: Expanding Horizons.* 2nd Edition. The Fairmont Press. Lilburn, Georgia USA.
_____ 2005. *Investment Grade Energy Audits: Making Smart Energy Choices.* The Fairmont Press. Lilburn, Georgia USA
Light's Labours Lost. Fact Sheet. International Energy Agency, 2006

Chapters 2 and 3. Western and Eastern Europe
Administration of Seversk. 2006. Program of the regional development for administration unit of Seversk 2006-2009. Annex II. Project "Providing heat to municipalities." The analysis of projects under investments of international financial organizations (in Russian).
Agence de l'Environnement et de la Maîtrise de l'Energie (ADEME). 2006. Current situation of the energy efficiency services market in France. Country overview. EUROCONTRACT project.
Aidonis, A. and G. Markoginnakis, 2006. Development of Pilot Solar Thermal Energy Service Companies (ST-ESCOs) with High Replication Potential. ST-ESCOs Market Analysis: Hellas. (Project Report of no. EIE/04/059/S07.38622). Albanian-EU Energy Efficiency Centre (AEEC) n.d. webpage. URL: http://www.eec.org.al [consulted 10 November 2006].
_____. n.d.b. Countries: Serbia and Montenegro. URL: http://www.ase.org/section/country/serbmont [consulted 19 December 2006].

Associazione Imprese di Facility Management ed Energia (AGESI). n.d. website. URL: www.agesi.it (partially in Italian).

Austrian Energy Agency (E.V.A.). 2005. Country Overview. EUROCONTRACT project.

BerliNews 17 May 2005. European Energy Service Award 2005. based on information from Berliner Energieagentur GmbH, Andrea Köhnen. (in German) URL: http://www.berlinews.de/archiv-2004/3446.shtml [consulted 5 August 2006].

Bertoldi, P., M. Hinnells, and S. Rezessy, 2006a. Liberating the power of energy services and ESCOs in a liberalised energy market. In: *Proceeding of the International Energy Efficient Domestic Appliances and Lighting Conference (EEDAL`06), London, 21-23 June 2006*. Eds: Bertoldi, P., Kiss, B., Atanasiu, B. Ispra, Italy: European Commission, DG Joint Research Center.

Bertoldi, P., S. Rezessy, and E. Vine, 2006b. Energy service companies in European countries: Current status and a strategy to foster their development. *Energy Policy* 34: 1818-1832.

Better Integration of Sustainable Energy (BISE). 2005. Reports by Countries: Development of Municipal Energy Efficiency Networking Activities. URL: http://www.bise-europe.org/IMG/pdf/National_reports_Bise.pdf [consulted 5August 2006].

Center for Renewable Energy Sources (CRES). 2005a. EPC in Greece: Current Situation. Country Overview. EUROCONTRACT project.

_____. 2005b. ST-ESCOs newsletter. Issue 4. URL: http://www.stescos.org/index.htm[consulted 28 August 2006].

Ceresi, G. 2005. Role of ESCO in the industrial marketing in Italy: Siram experience. Presentation at *ESCO Europe Conference 2005*. 4-5 October 2005, Vienna.

Chabchoub, J. 2005. Country Summaries (Part 2) The Environment for Energy Performance Contracting in Central Europe. Monthly Balkan Energy Solutions Team (BEST) e-mail bulletin in power systems, renewable energy sources, electricity market and ecology 16: 9-15.

Chistyakova, O.N., Allen Morin, and A. Pasoyan, 2006. *Removing Barriers to Residential Energy Efficiency in Southeast Europe and the Commonwealth of Independent States*. Kiev, Ukraine: Alliance to Save Energy.

De Groote, W. 2006. ESCO`s for households: A New Phenomena in Europe? In: *Proceeding of the International Energy Efficient Domestic Appliances and Lighting Conference (EEDAL`06), London, 21-23 June 2006*. Ispra, Italy: European Commission, DG Joint Research Center.

Energikontor Sydost. 2005. *EPC in Sweden.* EUROCONTRACT project.

Energy Center Bratislava (ECB), n.d.a., Framework Conditions for Energy Performance Contracting and Delivery Contracting in Public Buildings—Slovakia (CLEARCONTRACT project). Bratislava, Slovakia: ECB.

Energy Center Bratislava (ECB), n.d.b., Potentials for Energy Performance Contracting and Delivery Contracting in Public Buildings—Slovakia (CLEARCONTRACT project). Bratislava, Slovakia: ECB.

Estrela, A. 2004. Efficient street lighting: integration of information technologies in energy management. In: *Proceedings of First European Conference of Municipal Energy Managers.* Stuttgart, Germany, 1-2 July 2004.

EU-Russia Energy Dialogue Technology Centre. 2006. Summary of the the Seminar on ESCOs and Gas Flaring In the Framework of the EU-Russia Energy Dialogue Moscow, Russia, 26 October 2006.

European Bank of Reconstruction and Development (EBRD). 1998. EBRD and EU encourage energy saving in Ukrainian small and medium-sized enterprises through loan to country's first energy service company (ESCO). EBRD Press Release 24 May 2006. URL: http://www.ebrd.com/new/pressrel/1998/24may9.htm [consulted on 10 December 2006].

European Commission, DG Joint Research Center (EC DG JRC). 2005. *European Energy Service Companies Status Report 2005.* Authors: Bertoldi, P. and Rezessy, S. Ispra, Italy: EC DG JRC.

Fanjek, J. and B. Šteko, 2005. Energy efficiency project in Croatia. Presentation at *ESCO Europe Conference 2005.* 4-5 October 2005, Vienna.

Forsberg, A., C. Lopes, and E. Öfverholm forthcoming. How to kick start a market for EPC—Lessons learned from a mix of measures in Sweden. In: *Proceedings of the European Council for Energy Efficient Economy 2007 Summer Study.* Stockholm: European Council for an Energy-Efficient Economy.

Geissler, M. 2005. EUROCONTRACT—Guaranteed Energy Performance. Standardised Energy Services for Europe's buildings. Presentation at ESCO Europe Conference 2005. 4-5 October 2005, Vienna.

Geissler, M., A. Waldmann, and R. Goldmann, 2006. Market development for energy services in the European Union. In: *2006 ACEEE Summer Study on Energy Efficiency in Buildings—"Less is More: En Route to Zero Energy Buildings."*

Grim, M. 2006. The Austrian programme for private service buildings: ecofacility. In: Proceedings of International Conference on Improving Energy Efficiency in Commercial Buildings (IEECB'06), Frankfurt (Germany), 26-27 April 2006. Eds. Bertoldi, P. and Atanasiu, B. Ispra, Italy: European Commission, DG Joint Research Center.

Hinnells, M. 2006. Aiming at a 60% reduction in CO_2: implications for residential lights and appliances and micro-generation. In Proceeding of the International Energy Efficient Domestic Appliances and Lighting Conference (EEDAL`06), London, 21-23 June 2006. Eds: Bertoldi, P., Kiss, B., Atanasiu, B. Ispra, Italy: European Commission, DG Joint Research Center.

Hyponnen, S. 2006. Boosting efficiency with ESCO service. Presentation at the European Conference on Developing the Energy Efficiency Market (DEEM). 21- 22 September 2006, Budapest.

Instituto para la Diversificación y Ahorro de la Energía (IDAE). n.d. webpage. URL: www.idae.es (in Spanish) [consulted 16 July 2006].

_____. 2005. *Energy Policies of IEA Countries: Belgium 2005 Review.* Paris: OECD/IEA.

Irrek, W., S. Attali, G. Benke, N. Borg, A. Figorski, M. Filipowicz, A. Ochoa, A. Pindar, and S. Thomas, 2005. PICO Light project, SAVE Contract No. 4.1031/Z/02-038/2002—Final Report. Döppersberg, Germany: Wuppertal Institut.

Irrek, W., S. Thomas, and G. Benke, 2006. Internal performance commitments enabling a continuous flow of energy efficiency measures. In: *Proceedings of International Conference on Improving Energy Efficiency in Commercial Buildings (IEECB'06),* Frankfurt (Germany), 26-27 April 2006. Eds. Bertoldi, P. and Atanasiu, B. Ispra, Italy: European Commission, DG Joint Research Center.

Ketting, J. 2006. Energy Efficiency in Russia: A Chance to Excel or a Hard Lesson to Learn? *Russia Investment Review* 4: 94-95.

MOTIVA Oy. 2005. Country Overview: Finland. EUROCONTRACT project.

MOTIVA Oy. n.d.. website. URL: www.motiva.fi (information on ESCOs is in Finish). [consulted 30 January 2007].

Murajda, T. 2005. Energy efficiency contract in district heating domain—elementary schools in Petrzalka by C-TERM spol. s.r.o. In: *Proceedings of the Energy Efficiency Potential in Buildings, Barriers and Ways to Finance Projects in New Member States and Candidate Countries.* Tallin, Estonia July 2005. Eds: Paolo Bertoldi and Bogdan Atanasiu. Ispra, Italy: European Commission, DG Joint Research Center.

MURE-Odyssee. 2006a. *Energy Efficiency Profile: Luxembourg.* Also available on-line: www.mure2.com.

MURE-Odyssee. 2006b. *Energy Efficiency Profile: Spain*. Also available online: www.mure2.com.
Pujol, T. 2004. The Barcelona solar thermal ordinance. In: *Proceedings of Annual Conference of Energie-Cités: Working in Synergy with the Private Sector?* Martigny, Switzerland, 22-23 April 2004.
Racolta, S. 2005. The UNDP/GEF Energy Efficiency Financing Team in Romania. In: *Proceedings of the Energy Efficiency Potential in Buildings, Barriers and Ways to Finance Projects in New Member States and Candidate Countries*. Tallin, Estonia July 2005. Eds: Paolo Bertoldi and Bogdan Atanasiu. Ispra, Italy: European Commission, DG Joint Research Center.
Rezessy, S., K. Dimitrov, D. Urge-Vorsatz, and S. Baruch, S. 2006. Municipalities and energy efficiency in countries in transition. Review of factors that determine municipal involvement in the markets for energy services and energy efficient equipment, or how to augment the role of municipalities as market players. *Energy Policy* 34(2): 223-237.
Rodics, G. 2005. ESCOs in the Hungarian Energy Market. In: *Proceedings of the Energy Efficiency Potential in Buildings, Barriers and Ways to Finance Projects in New Member States and Candidate Countries*. Tallin, Estonia July 2005. Eds: Paolo Bertoldi and Bogdan Atanasiu. Ispra, Italy: European Commission, DG Joint Research Center.
Russian Energy Efficiency Demonstration Zones (Rusdem). n.d. website. URL: http://www.rusdem.com/Pages/index.htm [consulted 17 November 2006].
Saffet Bora, F. 2007. A New Era in Energy Efficiency in Turkey. Energy Review 9:2-4. URL:http://www.turkishweekly.net/energyreview/TurkishWeekly-EnergyReview 9.pdf [consulted 5 March 2007].
Sehovic, H. 2005b. BiH Experience in Energy Efficiency, Energy Efficiency Financing. Presentation at the Energy Efficiency Investment for Climate Change Mitigation.
Sorrel, S. 2005. *The Contribution of Energy Services Contracting to a Low Carbon Economy*. Tyndall Centre Working Paper, Environment & Energy Programme SPRU (Science & Technology Policy Research), Freeman Centre.
ST-ESCO project. 2006a. ST-ESCOs Market Analysis: Austria. Project Document. Project no. EIE/04/059/S07.38622.
ST-ESCO project. 2006b. ST-ESCOs Market Analysis: Spain. Project Document. Project no. EIE/04/059/S07.38622.
USAID. 2005. Credit Guarantees Promoting Private Investment in Devel-

opment. Year Review 2005. Washington, DC: USAID.
Vegel, M. 2006. *Eurocontract. European Platform for the Promotion of Energy Performance Contracting.* Presentation at the ESCO Europe 2006 International Conference, Prague, 26-27. September 2006.
Vine, E. 2005. An international survey of the energy service company (ESCO) industry. *Energy Policy* 33: 691-704.
Zachariev, D. 2005. ESCO in Bulgaria: Projects, market, barriers. In: *Proceedings of the Energy Efficiency Potential in Buildings, Barriers and Ways to Finance Projects in New Member States and Candidate Countries.* Tallin, Estonia July 2005. Eds:
Paolo Bertoldi and Bogdan Atanasiu. Ispra, Italy: European Commission, DG Joint Research Center.
Zeman, J. 2005. Public tenders for EPC. Presentation at *ESCO Europe Conference 2005.* 4-5 October 2005, Vienna.
Zeman, J. and B. Dasek, 2005. ESCO in Czech Republic: projects, market, barrier. In: *Proceedings of the Energy Efficiency Potential in Buildings, Barriers and Ways to Finance Projects in New Member States and Candidate Countries.* Tallin, Estonia 100 July 2005. Eds: Paolo Bertoldi and Bogdan Atanasiu. Ispra, Italy: European Commission, DG Joint Research Center.
Žídek, O. 2005. Energy Performance Contracting in the Czech Republic—history, present and future development. Presentation at ESCO Europe Conference 2005. 4-5 October 2005, Vienna

Chapter 4. Africa
South Africa
Dieter Krueger, Crown Publications. *Energy Efficiency Made Simple.* 2006.
Eskom. Eskom DSM website.
Eskom. Eskom CTAD website.
SAAE. SAAE website.
Albert Africa. *DSM: Coming of Age in South Africa.* 2003.

Chapter 6. Asia
China
EMCA annual report (2007)
ESCO industry survey report (2007)
Taylor, R.P., Govindarajalu, C., Levin, J., Meyer, A.S., Ward, W.A.: *Financing Energy Efficiency: Lessons from Brazil, China, India and Beyond*; ESMAP/World Bank, 2008.

India

Taylor, R.P., Govindarajalu, C., Levin, J., Meyer, A.S., Ward, W.A.: *Financing Energy Efficiency: Lessons from Brazil, China, India and Beyond*; ESMAP/World Bank, 2008.

Japan

MURAOKSHI Chiharu, Hidetoshi NAKAGAMI, Tsuyoshi SUMIZAWA, The Investigation research on evaluation of a ESCO experimental project, in proceedings of the 16th conference on Energy, Economy and Environment, Japan Society of Energy and Resources, 2000.1

MURAKOSHI, Chiharu, Hidetoshi NAKAGAMI, Tsuyoshi SUMIZAWA: Exploring the feasibility of ESCO business in Japan—demonstration by experimental study, in Proceedings of the ACEEE 2000 Summer Study on Energy Efficiency in Buildings, 2000.8

New construction subcommittee: International Performance Measurement & Verification Protocol Volume 3, Efficiency Valuation Organization (EVO), 2006.1.

Measurement and Verification Protocol Committee: Study for the Measurement and Verification Protocol of the energy-saving effect, Energy Conservation Center Japan, 2001.3

Measurement and Verification Protocol Study Committee: The Measurement and Verification Protocol Guideline of the energy-saving effect, Energy Conservation Center Japan, 2002.3

Jyukankyo research institute: Basic investigation of guideline policy in Eco-Energy City Osaka, report, Osaka, 2001.3

Comprehensive resources energy board-of-inquiry energy-saving committee: Energy-saving committee report -about the state of the future measure against energy saving, 2001.6

Energy Conservation Center Japan: The investigation report for the ESCO industrial spread in local authorities, 2004.2

Energy Conservation Center Japan: Investigation projects about superior ESCO project commendation system examination, 2005.10

Ministry of Environment: The statement-of-principles description data about promotion of the contract which considered curtailment of emission of greenhouse gas in the nation, independent administrative agencies, etc., 2007.12

JAESCO survey 2007

Jyukankyo Research Institute: Heisei 15 fiscal year. The PFI practical use ESCO project introduction to governmental facilities, Ministry of

Economy, Trade and Industry, 2004.3

MURAKOSHI, Chiharu, Toshiyuki WATANABE, Yasunori AKASHI and Hidetoshi NAKAGAMI: Study on the development circumstances and the characteristics of ESCO business in Japan, J.of Architecture and Urban Design, Kyusyu University No.12, P91-101, 2007.7

MURAKOSHI, Chiharu, Toshiyuki WATANABE, Yasunori AKASHI and Hidetoshi NAKAGAMI: Study on the characteristics of ESCO business in Japan, Architectural Institute of Japan, Journal of Environmental Engineering No,622, 2008.2

Philippines

Department of Energy. 2006. *Philippine Energy Plan: 2006 Update.* Manila, Philippines.

Herrera, Alice B., Ph.D. 2005. *Energy Efficiency Opportunities and Investment Requirements in the Philippines.* A Report Prepared for the Asian Development Bank.

Marquez, Raymond A. (2005). *Comparative Analysis of ASEAN ESCOs.* A Report prepared for the UNDP-GEF-DOE Philippine Efficient Lighting Market Transformation (PELMAT) Project.

Marquez, Raymond A. (2006). *Output Report on Assistance Provided for DBP Model ESCO Transaction Project and BDO Demo Project.* A Report prepared for the UNDP-GEF-DOE Philippine Efficient Lighting Market Transformation (PELMAT) Project.

Marquez, Raymond A. (2007). *ESCO Framework of Cooperation.*

Ver, Antonio A. (2007). *ESCO Association of the Philippines.* Presented to the 2nd Asia ESCO Symposium, Tokyo, Japan. February 1-2, 2007.

Chapter 7. North America
United States

Cudahy, R.D. & T.K. Dreessen: *A Review of the Energy Service Company (ESCO) Industry in the United States,* prepared for the Industry and Energy Department, The World Bank, Washington D.C., March 1996.

Canada

Bonfils, Sibi; Pierre Langlois, and Gaby Polissois, (1997) *Energy Efficiency Projects and their Financing Mechanisms,* IEPF, Collection Cahier Prisme, Econoler International.

Langlois, Pierre and Yves Robertson, (2003) *Demand-Side Management from a*

Sustainable Development Perspective—Experiences from Québec (Canada) and India, Econoler International Teri, IREDA, 2003

Mexico
Energy Efficiency Financing Assessment Report from IEEFP Mexico, EVO, ESP Capital and Mónica Pérez, May 2006
ACEEE Summer Study, Successfully Advancing Energy Conservation Efforts in Mexico, Rennè, Cohen, and Pérez Ortiz, 2004
International Energy Efficiency Financing Protocol IEEFP-Mexico-Bank Training Manual, EPS Capital, EVO and Mónica Pérez, January 2008
Thomas Dreessen EPS Capital, March 2008
ESMAP webpage: http://www.esmap.org/activities/index. asp?sort=title&s=80
Optima Energia, March 2008

Chapter 8. South America
Brazil
ABESCO (Coordinator A.D. Poole & M.C. Amaral): *Análise dos Resultados da Pesquisa das Empresas de Serviços de Eficiência Energética no Brasil*; prepared for the World Bank/UNEP/UNF program "New Financial Intermediation Mechanisms for Energy Efficiency Projects in Brazil, China and Índia," São Paulo, February 12, 2005.
ANEEL/SPE: *Manual para Elaboração do Programa de Eficiência Energética—2008*; prepared by the Superintendency for Reasearch & Development and Energy Efficiency; approved by Resolution #300 of February 12, 2008.
CTEE (Comitê Técnico para Eficientização do Uso de Energia): *Plano Energia Brasil—Eficiência Energética*; October, 2001.
Garcia, A.G.P.; *Leilão de eficiência energética no Brasil*; Doctoral thesis for the Federal University of Rio de Janeiro/COPPE, January, 2008
Lima, L.E.A.; C.M. Ayres, A.D. Poole, C.F. Hackerott, M. Campos: *Analysis of the Viability and Design of a Guarantee Facility for Energy Efficiency Projects*; prepared for the World Bank/UNEP/UNF program "New Financial Intermediation Mechanisms for Energy Efficiency Projects in Brazil, China and India," August, 2005.
Marçal, M.E. & P.C. Magalhães: *Opportunities and Challenges in the Development of Financial Intermediation Mechanisms for Energy Efficiency Projects in Brazil*; prepared for the World Bank/UNEP/UNF program "New Financial Intermediation Mechanisms for Energy Efficiency

Projects in Brazil, China and India," April, 2005.
Marçal, M.E.: *Considerations for Structuring a Trade Receivables Fund ("FIDC") to Finance Energy Efficiency Projects in Brazil*; prepared for the World Bank/UNEP/UNF program "New Financial Intermediation Mechanisms for Energy Efficiency Projects in Brazil, China and India," November, 2005.
Nexant: *Contratos de Desempenho para Serviços de Eficiência Energética no Setor Público do Brasil: Questões Jurídicas e Possíveis Soluções*; report to the Brazilian Ministry of Mines and Energy with support from USAID, January, 2004.
Poole, A.D. & A.S. Meyer; *Brazil Country Report*; prepared in English and Portuguese for the project "Developing Financial Intermediation Mechanisms for Energy Efficiency Projects in Brazil, China and India," World Bank, August 2006.
Poole, A.D. & J.B.N. Poole: *Summary of Results of the Survey of Brazilian Energy Efficiency Service Providers*; prepared for the CIDA program "Greenhouse Gas Emissions Reduction in Brazilian Industry" (GERBI), Rio de Janeiro, December, 2003.
Taylor, R.P., C. Govindarajalu, J. Levin, A.S. Meyer, W.A. Ward: *Financing Energy Efficiency: Lessons from Brazil, China, India and Beyond*; ESMAP/World Bank, 2008.

Chile

Comisión Nacional de Energía, CNE, (2004),"Estimating Potential Energy Savings by Improving the Energy Efficiency of Different Consumption Sectors in Chile."
Programa País de Eficiencia Energética 2006-2007.
Programa País de Eficiencia Energética (2005-2006) Energy Efficiency Directory of Chile.
Programa País de Eficiencia Energética Strategic Plan 2007-2015 of Programa País Eficiencia Energética de Chile.
Fundación Chile, (2007), Methodology to Identify and Assess Energy Services Companies (ESCO), final report within the framework of project BID-FOMIN, produced by Econoler International.
Fundación Chile (2007a), Study to Identify Market Potential and Approach, final report within the framework of project BID-Fomin, produced by Gamma Ingenieros S.A., Santiago.
Fundacion Chile (2007b), Proposal of Iinstruments and Sector Models for Clean Energy and Energy Efficiency, status report, Project Inter-

American Development Bank—FOMIN, Study conducted by Econoler International.

Uruguay
Energy Market and ESCO Market Assessment—Econergy 2002.
PAD—Project Appraisal Document—World Bank—Uruguay Energy Efficiency Project.
Balance Energético Nacional—2006—Dirección Nacional de Energía y Tecnología Nuclear—Ministerio de Industria, Energía y Minería Web page www.eficienciaenergetica.gub.uy.
Market study for UTE-USCO—ECONOLER International—2008.
Web page www.eficientlighting.net/doc/20070108(2).
UTE-USCO—M. González—2008.

Chapter 10. The Global Picture
Hansen, Shirley J. "Making the Business Case for Energy Efficiency," International Energy Efficiency in Commercial Buildings Conference. 2004. Frankfort, Germany. Paper also presented as keynote to annual conference of Japan Association of Energy Service Companies, 2005.

Appendix A
Contributing Authors

Many of our contributing authors were kind enough to supply short biographies, so our readers are able to have a better sense of the backgrounds and familiarity with performance contracting each author brings to the task. These biographies are presented below in chapter order and then alphabetically by country within the chapter.

CHAPTER 2. WESTERN EUROPE

Belgium

Lieven Vanstraelen started his career at Belgacom, Belgium's incumbent telecommunications company, where he worked for 4 years at the design and implementation of a new national transmission network. In 1996, he became one of the pioneers of Hermes Europe Railtel (later to become GTS/Ebone) which was the first private pan-European telecommunications carrier. In 2000, Vanstraelen moved to France where he was Marketing Director of the content technology provider ActiVia Networks.

He moved from telecommunication, to the energy efficiency sector in 2003. His focus was originally on online energy monitoring solutions. He became Managing Director, France, and Chief Marketing Officer of the Belgian energy services and monitoring solutions provider, REUS. In 2003 he co-founded "eden" the French-European network for innovation in the cleantech and energy industry. After his return to Belgium, he continued to represent "eden"—which later changed its name to "Agora Energy." Today the organization has a network of around 2,500 people in Europe.

He was appointed Managing Director of Fedesco, the Belgian public ESCO and Third Party Investor in December 2006. Fedesco focuses on energy savings and renewable energy in federal public buildings and know-how transfer to other public authorities.

Vanstraelen co-founded, and became member of the steering

committee of EMAB, the Energy Managers Association of Belgium. He also co-founded BELESCO, the Belgian ESCO Association and is currently a member of the program committee for the ESCO Europe conference.

France

Jérôme Adnot is a Full Professor at the Ecole des Mines de Paris (now Mines-Paristech). Since 1991, his main research theme has been the evaluation of a number of new energy techniques. He is the co-author of numerous contractual reports on the subject, namely for SAVE (CEC-DGTREN in Brussels) and ADEME, the French environmental and energy management agency. From 1997 to 1999 he was coordinator of EERAC: "Energy Efficiency and Certification of Central Air Conditioners" and from 2000 to 2002 coordinator of "EECCAC: Energy Efficiency a of Room Air Conditioners" (both for DGTREN, first about small systems, then the study of large systems).

Frédéric Rosenstein, an engineer, is responsible for the energy efficiency services and integrated resources planning activities of the DSM division at ADEME, the French environmental and energy management agency.

United Kingdom

Anees Iqbal is the head of Maicon Associates Ltd., a consultancy dedicated to providing technology transfer and training services in energy management and ESCO development.

He has over 40 years experience in the energy industry, with more than 20 years in performance contracting. He was a director at Emstar, an energy management subsidiary of Shell UK and the first ESCO in UK. This later became part of Dalkia, the utility arm of the French Multi-national Group, Vivendi.

Anees and his Maicon team now provide consultancy and ESCO technology transfer services to clients which include the EU, various national governments, and IFIs such as the World Bank. Anees has had an extensive involvement in promotion of ESCOs in the Czech Republic, Poland, Lithuania, Belarus and Romania.

He holds bachelors and master's degrees in engineering (Southampton University). He is a Professional Chartered Engineer, has spoken at many international conferences and has presented and published numerous technical articles including a text book on noise control, and contributed to a chapter in a book on performance contracting, by Dr. Shirley Hansen in the United States.

European Commission
Paolo Bertoldi, DG, The Joint Research Center for the European Commission, is the co-author of the report, *Latest Development of Energy Servicer Companies across Europe—A European ESCO Update,* published in 2007, upon which portions of Chapters 2 are based.

CHAPTER 3. EASTERN EUROPE
Bulgaria
Pavel Manchev was born in 1949 in Sofia, Bulgaria. He has a master's degree from the University of National and World Economy in Sofia. His expertise in software development and systems design led him to become the director of the department for software development in Software Products and Systems Corporation (SPS). In the late 80s, Manchev was a representative for SPS in Prague and later the deputy director of TESOS—Bulgaria. In1994, he joined the Center for Energy Efficiency, where he served as Managing Director of its subsidiary, performing energy auditing, financial engineering and consultations in energy efficiency and renewable energy. He is an expert in managing projects and has done so for USAID, GEF, European Commission, UNDP, UNECE, JICA and the World Bank.

Poland
Janusz Mazur, since 2005, has been president of Przedsiębiorstwo Oszczędzania Energii ESCO Ltd. in Krakow, Poland (POE ESCO—Energy Saving Company). This company was created in 2000 by the Municipal District Heating Company (MPEC SA) in Krakow, with World Bank assistance. Prior to POE ESCO, Janusz Mazur was the head of the department of Strategy and Promotion in the MPEC SA—the owner of POE ESCO.

He has a Master of Science degree in mechanical engineering, from the University of Mining and Metallurgy and The Krakow School of Business at the University of Economics in Krakow. He focuses his interests around saving electric and thermal energy. Recently, he has prepared projects to finance, in the ESCO model, and changes of bus fleets driven with diesel to be driven with compressed natural gas (CNG).

Mazur has spoken at many domestic and international conferences on energy conservation and related concerns. He also has published in this area. He has cooperated with many international organizations, including The World Bank, GEF, European Bank of Reconstruction and Development, US Trade and Development Agency, USAID, US Department of Energy and the Joint Research Center European of the European Commission. He is the proud recipient of the Golden Badge of the Polish District Heating Industry Chamber.

Russia

Igor Rokhlikov is one of the first national private consultants who initiated interest in ESCOs in Russia in the late 90s. Since that time, he has coordinated eight international projects of inter-regional scope in which experts and companies from several Europeans countries participated.

His company, Regional Economic Development Agency, is well known in the national and international business environment for its practical results in energy efficiency and energy saving in industries and in the public sector.

European Commission

Paolo Bertoldi, DG, The Joint Research Center for the European Commission. He is the co-author of the report, *Latest Development of Energy Servicer Companies across Europe—A European ESCO Update,* published in 2007, upon which portions of chapter 3 are based.

CHAPTER 4. AFRICA

Algeria, Morocco, and Tunisia

Hakim Zahar is a senior energy efficiency expert with over 30 years of experience in North African countries, as well as in several additional countries around the world. He has been instrumental in the start of the first ESCO operation in Tunisia, as a subsidiary of Econoler International. For the last 10 years, he has acted a vice president for Econoler, a well renowned international consulting firm in the energy efficiency and clean energy sectors.

His international experience lead him to work on an impressive number of projects carried out or financed by a number of international institutions, the United Nations and with bilateral or multilateral development organizations.

Côte d'Ivoire

Dr. M'Gbra N'Guessan is one of the most recognized clean energy experts in the West Africa region. He has more than 20 years of experience in energy efficiency and renewable energy in all the region, as well as in North Africa and in many other countries around the world. Between 1996 and 1998, he was the coordinator of a major energy efficiency project for the United Nations Development Program (UNDP) and acted as Environment Director of Côte d'Ivoire in 1999.

Since 2000, he has served as Vice President for Africa for Econoler International. His international experience lead him to work on an impressive number of projects carried out or financed by a number of international institutions, the United Nations and with bilateral or multilateral development organizations.

Kenya

Paul Kirai is an energy management and environmental management specialist. He has expertise in energy management, energy efficiency, energy standards and labeling, climate change and CDM. Has worked with governments, the United Nations, the private sector and regional organizations. In 2001-2006, he was the National Project Manager for the GEF-KAM Energy Efficiency Project in Kenya funded by GEF. The project created impact at industry and government levels and is hailed as a good best practice by UNDP and GEF for its efforts in mainstreaming energy efficiency in industry and national institutions. He has been a speaker in many energy forums and has authored a number of publications on energy as well as on environment and climate change.

He is a member of the national climate change action committee, and serves as Chair—Mitigation and Industry. He is also a member of the Association of Energy Engineers, and holds certificates in energy management, energy auditing and financial engineering.

South Africa

Braam Dalgleish is an engineer employed by Energy Cybernetics (Pty)Ltd and has been involved in the energy field for the past nine years. He is certified by the Association of Energy Engineers (AEE) as a certified energy manager. He has been actively involved in delivering energy training, energy management, measurement and verification, and energy cost saving services to industry.

Dalgleish is currently enrolled for this Ph.D. in mechanical engineering at the North-West University, South Africa. He has authored and co-authored more than 10 local and international articles on energy related topics. He has also presented and co-presented more than 15 papers at local and international conferences and seminars on energy topics. He has conducted more than 50 industrial and commercial energy audits.

Dr. L.J. Grobler is co-director of the company Energy Cybernetics CC. He holds CEM and CMVP credentials from the Association of Energy Engineers (AEE). He is registered as a professional engineer (Pr Eng) with the Engineering Council of South Africa and received his Ph.D. at the University of Pretoria, South Africa.

He is president of AEE and chapter president of the Southern African Chapter of AEE. He administers the CEM and CMVP training programs for the AEE in Southern Africa.

LJ is also a professor in mechanical engineering at the School for Mechanical Engineering at the NorthWest University. He has authored and co-authored more than 40 local and international articles on energy related topics. He has also presented and co-presented more than 50 papers at local and international conferences and seminars on energy topics. He has conducted more than 100 industrial and commercial energy audits.

CHAPTER 5. ESCOs IN THE MIDDLE EAST

Egypt

Emad Hassan is a principal with Nexant, Inc. (a U.S. consulting firm), with 25 years of experience in demand-side energy management, in the U.S. and internationally. He is managing a global energy efficiency initiative operating in five countries on three continents with a focus on energy access to developing markets. In the early 90's, he headed the operations of one of the most pioneering and successful utility-sponsored ESCO initiatives in Southern California. In Egypt, he was instrumental in the formation of the first energy service NGO in 1999 (the Egyptian Energy Service Business Association—EESBA). He served on the board of National Association of Energy Service Companies and received his master's degree from UCLA's School of Architecture and Urban Planning.

Israel

Z'ev Gross has been active in the field of energy efficiency since 2003. He holds academic degrees in both the natural sciences (geology) and law. He practiced corporate law till 2002, when he joined the Israel Ministry of

National Infrastructures.

As the legal counsel to the Infrastructure Resources Management Division of the Ministry, which is responsible for energy efficiency, Z'ev was instrumental in reviving the regulatory activity of the division and, with the assistance of Econoler International Inc., in introducing the concept of performance contracting in the energy efficiency field in Israel. In 2004, he became acting head of the division and in 2006 was appointed to the post officially.

Z'ev was instrumental in rewriting the energy efficiency agenda in Israel, articulating, for the first time, the two pronged approach of Market Transformation and Resource Acquisition, as well as promoting the various tools that support each of these two different types of activity. The culmination of his activities to date has been the resolution adopted by the Israeli Government for the establishment of a national energy efficiency program. In addition to contributing to articulation of the program in general, Z'ev continues to work towards the completion of the various tools needed to support the program—a revolving loan fund, tax incentives, a residential sector oriented DSM program suite, and various other tools of a market-oriented nature.

In addition to dealing with energy efficiency, Z'ev's corporate background has been utilized in helping promote projects in renewable energy, as well as promoting R&D activities in the area of renewables.

Turkey

Larry Good is an international energy project developer, working abroad since 1999 in the former Soviet republics, eastern Europe, Brazil, India, Cyprus and Turkey. He has 26 years of engineering experience and 25 years of management experience. He currently serves as vice president of Envo Energy services. In the 90s, he conducted energy audits at federal facilities in the US Postal Service. He wrote the technical specification for an USD 11 million energy performance contract that reduced energy consumption and costs by 60 percent in the US Environmental Protection Agency's national auto testing laboratory. He currently manages and markets energy services for Envo Energy in Turkey. He is a past president of the Association of Energy Engineers (AEE) and currently serves as the director of AEE's International Certification Board.

Dr. Halil Guven's doctoral work was in mechanical engineering, focused on energy and the computer aided design of thermal energy systems. He conducted a project funded by USAID and the government of India to investigate technology transfer issues. He has worked as an instructor, associate professor and professor at various universities. He has also served as a director or rector of several centers that address energy-related and sustainable environmental management plans. He has over 50 publications to his name and many energy professional development and energy innovation awards.

Ata Osman Memik has over 15 years of professional business and management experience in industry and consulting. He has a BS degree in finance. From 1993 to 1997 he worked as a budget executive & deputy financial coordinator of Penguen Foods ($40 M turnover). He established the budget department, improved cost analysis and production planning, worked with the banking consortiums as the key contact to prepare all necessary analysis and reports to enter the stock market. From 1997 to 1999 he worked for Tusan Foods ($60 M turnover). There he served as the budget & finance manager and helped the company to restructure loans, maintain a baseline to open into the stock market and establish the budget and financial control departments in five production plants. Between the years of 1999 and 2004 he worked as the general manager for National Britannia, Turkey (food safety & environmental risk management service, auditing and consultancy). He was honored with the first International Britannia Man of the Year award as the best general manager among the 20 countries under the International Britannia Group. From 2004 to the present he is the general manager and partner of Envo Group Ltd, International Trade and Consulting. He set up the transit trading department, increased the turnover by $10 M and set up the infrastructure to get the telecom license for long distance calls. Currently Osman is focused on building up the first ESCO in Turkey with energy management consulting & project services.

CHAPTER 6. ASIA

China

Zhao, Ming (Lily) obtained her masters degree in environmental management for business from Cranfield University in UK. She has been working on environment and energy consulting for many years in world famous consulting companies. After she returned from UK, she joined EMCA-China Energy Conservation Association Energy Service Industry Committee in China, working as vice director and secretary general. EMCA is a national ESCO association with more than 300 members from across the country. By providing effective and efficient service to the members and co-operation with government agencies, international organizations and relevant research institution, EMCA plays a more and more important and active role in promoting EPC (energy performance contracting) and the ESCO business in China.

Shen Longhai is a senior engineer. He studied in Harvard Business School (AMP104) in the U.S.A. He has been engaged in energy development, energy conservation and environmental protection for a long time. He formerly served as Director General, Dept. of Energy of the State Economic Commission; Dept. of Resources Conservation and Comprehensive Utilization, Dept. of Spatial Planning and Regional Economy of the State Development Planning Commission; Vice Chairman of China Energy of the State Development Planning Commission; Vice Chairman of China Energy Research Society; Economic Counselor of Chinese Embassy in the U.S.A. He now serves as the director of EMCA, China's association for energy service companies.

India

Nisha Menon is the senior consultant to DSCL Energy Services Company. Prior to joining DSCL, Ms Menon worked at IREDA and Winrock International, and provided the information on private sector activities in India. She holds a postgraduate degree in Energy Conservation and Management with specialization in energy efficiency and renewable energy. She has 15 years of professional experience and spearheads the energy strategy and management consulting business of DSCLES. Her contribution to the book includes inputs from DSCL.

D.V. Satya Kumar provided a study conducted by Econoler International in 2007 on public sector activities in India and related comments. Mr. Kumar is managing director of Shri Shakti Alternative Energy Limited.

Japan

Dr. Chiharu Murakoshi is the vice president of Jyukankyo Research Institute, Inc. and Secretary General of the Japan Association of Energy Service Companies (JAESCO). He has done investigative research on energy demand analysis of the residential and commercial sector, including energy consumption analysis, energy saving, alternative energy technology, energy efficiency policy, measures against global warming and research on the ESCO industry. Themes of his research include database development and demand analysis. He has also helped promote energy efficiency policy and the ESCO industry.

Murakoshi has served as a member of the energy council of the Ministry of Economy, Trade and Industry. He took charge of the secretariat of JAESCO in 1999 and planned and managed the 1st Asia ESCO conference in 2005 and the 2nd Asia ESCO conference in 2007.

Dr. Hidetoshi Nakagami founded the Jyukankyo Research Institute, Inc. in 1973 and has served as its president from that time to the present. He is a certified and authorized first class architect and builder and a registered consultant of the World Bank. He regularly lecturers on energy situations and strategic options in Japan and has served as professor at Tokyo Institute of Technology and at Keio University. He has also served as a guest professor at Waseda University. He served in the early 1990s on the Energy Efficiency Standard for Appliances in the Ministry of International Trade and Industry and as chairman of Implementation Study Committee for Daylight Saving, Energy Conservation Center of Japan. He currently serves on the Advisory Committee for Energy Policy, Ministry of International Trade and Industry, the committee for Energy Consumption Labeling of Appliances for the Energy Conservation Center and the Study Committee of Technical Countermeasure for Global Warming Prevention, Environment Agency of Japan. He is the Vice chairman of JAESCO

Takashi Masuda is a senior researcher at Jyukakyo Research Institute, Inc. He previously worked at Sankyu Incorporated and was a visiting researcher at the Institute of Energy Economics. Since 2000, a focus of his work at JRI has been the ESCO business. He has been secretary of JAESCO since 2003.

Korea

Ed Sugay is the global director of Energy and Environmental Solutions (EES) within Siemens Building Technologies. He is responsible for developing Siemens' EES business in Australia, China and Korea. Prior to working with Siemens, he was a project develop for a multi-national energy company, where he developed generation-scale power projects in AP as well as CHP projects for industrial customers.

Malaysia

Dr. Jim K. Y Lim is a pioneer in the development of Software and Hardware in Malaysia using information and communication technology

(ICT) in the monitoring and management of "Energy, the, REAL TIME ONLINE, Information Solution." He is a system designer, integrator and analyst in the field of energy management. He was a member of the Jabatan Bekalan Elektrik Technical Advisory Council (TAC) in 2002

Currently he is very involved in research and development for the Human Capital Development Programme, in many sectors of industry. He is a key speaker for the Energy Commission of Malaysia, has presented a paper to the Institute of Engineers of Malaysia, and many other international and national seminars on energy efficiency.

Philippines

Dr. Alice B. Herrera served as an energy consultant in the Philippines from 2001 to 2007 before migrating to Canada in 2007. She has conducted several policy and technical studies on renewable energy and energy efficiency with the main goal of promoting sustainable and clean energy. She was the president of the Energy Efficiency Practitioners Association of the Philippines, Inc., a non-stock, non-profit professional organization of energy managers and engineers. She also served as a supervising science research specialist at the Industrial Technology Development Institute, a government research agency under the Philippine Department of Science and Technology.

Thailand

Arthit Vechakij is managing director of Excellent Energy International Co., Ltd. (EEI). He pioneered Thailand's ESCO business in 1999 by making EEI a selected pilot ESCO of the World Bank/EGAT ESCO Pilot Project of Thailand. He has made important contributions to the development of the company and to the ESCO industry. He is well recognized as an expert in energy efficiency and ESCO areas by local and international agencies; government, alliances, media, and industrialists through being a public speaker at many energy events in the past nine years.

Ruamlarp Anantasanta serves as the deputy managing director—Marketing & Business Development, for EEI. He has worked in the project management and consulting business since the beginning of his career. Before joining EEI, he worked with Accenture Solutions, Inc. He joined EEI in 2004 to further extend its growth and maintain its leading position among Thailand ESCOs by strengthening the EEI brand and continuously acquiring new ESCO projects. He is recognized as a marketer and business developer in energy efficiency and the ESCO business by many local and international agencies.

Vietnam

Ha Dang Son is the deputy director of the Research Center in Energy and Environment (RCEE) of Vietnam, and an expert in renewable energy and energy efficiency. He is a graduate engineer with Master's degrees in Thermal Engineering (awarded by Hanoi University of Technology), and in Industrial Engineering & Management (awarded by the Asian Institute of Technology).

Ha Dang Son has extensive and practical experience in energy efficiency and ESCO development. He is currently the project developer and training manager of the joint project entitled "Vietaudit phase 2" funded by the Finnish Ministry of Foreign Affairs and the Ministry of Industry and Trade of Vietnam, aiming to promote ESCO business in Vietnam through on-the-job training activities. He is also involved in various projects providing energy services to cement and garment companies in Vietnam.

CHAPTER 7. NORTH AMERICA

Canada

Pierre Langlois is a senior expert with over 20 years of experience in the energy sector and more specifically in the development and implementation of energy efficiency and renewable energy projects and programs in Canada and internationally. He is recognized as one of the leading international experts in the development and implementation of innovative energy efficiency financial mechanisms and in the start-up and operation of Energy Service Companies (ESCOs) in industrialized countries and in transition or developing countries. He has also gained great expertise in the development of energy policies and demand-side management, as well as in the development of specific projects for international financial institutions.

His international experience has led him to work in more than 35 countries on projects carried out or financed by most major international institutions. During his career, he has acted as member of the board of directors for several ESCOs throughout the world and for some international organizations, such as the Efficiency Valuation Organization (EVO).

His work for Econoler has been recognized many times. He has received the AEE "Legend in Energy," a special lifetime recognition granted by the Association of Energy Engineers, as well as an award at the

Clean Technology Award 2000, given by the Climate Technology Initiative (led by the International Energy Agency) and the OECD at the Hague convention on climate changes (COP 6).

Mexico

Monica Perez is an architect from the Universidad La Salle Campus Mexico City, and holds a Master of Science in energy efficient buildings from Oxford Brookes University.

Perez was hired by the Efficiency Valuation Office (EVO) to participate in the development of the International Energy Efficiency Financing Protocol for Mexico (IEEFP-Mexico). She served as the liaison consultant in charge of communications with the Mexican National Commission of Energy Conservation (CONAE), the Ministry of Energy (SENER), and several other organizations which participate in the development of the Financing Protocol, such as commercial banks, project developers, support institutions, government sector, etc. She prepared Mexico's Energy Efficiency Financing Assessment Report, which was the foundation information for the development of the Finance Protocol in Mexico.

In parallel, she has worked for EPS Capital Corp to identify three demonstration projects that will integrate new innovative guarantee mechanisms that Nafin will provide in order to help the Mexican ESCO market grow and consolidate through a more active participation of local banks.

During 2000 to 2005, Perez worked for the US National Renewable Energy Laboratory (NREL) as the in-country coordinator of the collaboration program Climate Technology Partnership (CTP), serving as the liaison between NREL and CONAE for the development of the Mexican Energy Services Companies (ESCO) Market.

Perez has also been involved in CONAE's request for APEC funds and provided assistance to the US Department of Energy with the APEC projects to promote financing of EE and RE projects in Mexico. She participates in the organization of national and international seminars and conferences to promote the Mexican ESCO market, as well as participating as a speaker in these events. She has worked to promote innovative financial mechanisms with several State Ministries in order to introduce performance contracting in the federal sector, to design trust funds for ESCO projects, and other ways to alleviate the difficulties for market penetration.

United States

Donald Gilligan is president of the National Association of Energy Service Companies (NAESCO), an organization of about 75 companies that deliver about $5 billion of energy efficiency and renewable projects annually. He coordinates NAESCO's state advocacy activities, promoting energy efficiency and distributed generation in state legislative, regulatory and policy forums. He has worked in the energy efficiency industry for 30 years, as a consultant, entrepreneur, and state government official. He is a graduate of Harvard University.

CHAPTER 8. SOUTH AMERICA

Brazil

Alan Douglas Poole is an independent consultant on the efficient supply and use of energy in diverse sectors, as well as the environmental, economic and social consequences of energy policies and projects. Primary, but not exclusive, emphasis has been on developing countries. He was based in Brazil from November 1981 until June 2008. His areas of work have included; the regulatory framework for the energy sector, with emphasis on the pricing of electricity and gas to final consumers; the terms for connecting distributed generation to the electric grid; the efficacy of utility demand response programs and of public benefit wire-charges to promote energy efficiency and renewable energy.

He has also devoted time to the analysis of the relation between economic growth and energy use in general and in specific sectors; evaluation of the potential for improving energy efficiency in different sectors of the economy; and, identification of market barriers and ways to overcome them. His work also includes work in the development of ESCOs, including: innovations in project financing; assessment of risks; adaptation of performance contracts to local conditions; procedures to measure and verify energy savings; arbitration of conflicts; and, ways to open the public sector procurement of projects to rationalize energy and water use.

Chile

Dr. Karin Gauer has her doctorate in economics from the Free University of Berlin, Germany, and has served as assistant professor and scientific research assistant at the Institute for Environmental Economics of the Technical University, Berlin. She has also served as project manager for German Development Agencies (Carl Duisberg Gesellschaft, formerly CDG now InWent), Gesellschaft für Technische Zusammenarbeit (GTZ). Since 2004, she has been an independent consultant in the fields of environmental economics, living in Santiago Chile.

Uruguay

Marcelo Ulises Gonzalez Ferreira has served has a junior engineer, a distribution engineer team chief, a distribution planning manager and a USCO manager. He has worked for UTE, a state owned utility since 1980. Among his responsibilities for the utility, he has performed the definition of new distribution voltage levels, the definition of methodology for distribution of load forecast and technical loss determination.

CHAPTER 9. DOWN UNDER

Australia

Dr. Paul Bannister is managing director of Exergy Australia Pty Ltd, one of Australia's leading energy management consultancies. He has a PhD from the Australian National University in the field of solar thermal power. He has spent his entire professional career in the field of energy management, with a specific focus on commercial sector buildings in Australia and New Zealand, and has presented and published extensively in conferences around the world on energy efficiency related topics.

Bannister specializes in technical energy efficiency work, and has conducted energy audits for hundreds of sites ranging from small commercial offices to major hospitals and from university campuses to gold mines. He is also well known for his role as primary technical developer of the Australian Building Greenhouse Rating Scheme, a world-leading methodology for benchmarking the performance of office buildings. Paul also has extensive experience in the new buildings sector, having been involved in many development projects, ranging from 700m^2 eco-developments to 70,000m^2 office buildings.

His experience in the ESCO sector arises from having worked for two ESCOs in the development, investigation, implementation and monitoring of several energy performance contracts.

Appendix B

Note: The "Declaration" below is presented as released, with no changes except for the correction of a few spelling errors and more egregious grammatical mistakes, which did not affect the intent.

DECLARATION

INTERNATIONAL PARTNERSHIP FOR ENERGY EFFICIENCY COOPERATION IPEEC)

(Released June 9, 2008)

Considering:
  Declaration of the Gleneagles, St. Petersburg and Heiligendamm Summits, which emphasize the need for global cooperation in the field of **energy efficiency**.

Recognizing:
  1. That improving **energy** saving and **energy efficiency** is one of the quickest, greenest, and most cost-effective ways to address **energy** security, climate change, and ensuring economic growth;
  2. That a comprehensive process has been launched to enable the full, effective and sustained implementation of the UNFCCC through long-term cooperative action, now up to and beyond 2012, in order to reach an agreed outcome and adopt a decision at the COP 15;
  3. That all countries, both developed and developing, share common interests for improving their **energy efficiency** performance, there is an abundant potential for **international cooperation** among them and developed countries need to play an important role in **cooperation** with developing countries, accelerating dissemination and transfer of best practices and efficient technologies and capacity building in developing countries, which will contribute to improvement of **energy efficiency** at a global level;

4. That measures to increase **energy efficiency** can help achieve other objectives, such as reducing environmental pollution. These co-benefits can significantly increase the attractiveness of **energy efficiency** measures;
5. That many countries promote **energy efficiency** through nationally defined **energy efficiency** goals/objectives and action plans taking into account their technological and economic development;
6. That to improve **energy efficiency** it is necessary to understand fully the **energy** market situation, to identify key energy consuming sectors, to analyze the potential for improving **energy efficiency** in these sectors, and to take necessary measures to realize this potential;
7. That accelerating the market uptake and affordability of the best applicable **energy efficiency** technologies should be supported through incentives, and regulatory, market and voluntary measures, research and development of cost-effective **energy efficiency** technologies should be fostered, through development of public/private partnerships and through expanded **international cooperation**;
8. That green and **energy**-efficient public procurement can generate significant benefits for the uptake of the **energy**-efficient technologies, thereby stimulating competitiveness between companies, especially in those sectors where public purchases make up a relatively large share of total purchases;
9. That it is important and valuable to exchange information, experience and best practices on the most effective means for promoting **energy efficiency**, taking into account the activity developed among countries in the context of **international** organizations and agreements;
10. That **cooperation** among all countries that support the dissemination of know-how and technology on **energy efficiency** can effectively promote global energy efficiency;
11. That all reliable and high quality **energy** use data, statistics and information systems are perquisites for effectively setting goals/objectives, implementing national action plans, and evaluating policies and measures;
12. That efforts to inform, educate and persuade individuals and stakeholders about **energy efficiency** should be strengthened;
13. That public-private partnerships in, and across, key energy

consuming sectors can be instrumental for improving energy efficiency; and

14. That **energy efficiency** improvements require conditions that enable investment inflow, such as access to capital, stronger markets for **energy** services and market-based **energy** pricing.

Canada, the People's Republic of China, France, Germany, India, Italy, Japan, the Republic of Korea, the Russian Federation, the United Kingdom, the United States of America and the European Community, represented by the European Commission, united by their common interest in promoting energy efficiency and energy savings decide to establish an International Partnership for Energy Efficiency Cooperation and will take the necessary steps to this end as follows:

Objectives

The purpose of the International Partnership for Energy Efficiency Cooperation is to facilitate these actions that yield high energy efficiency gains. Participants in the Partnership choose to take action in the areas of their interest on a voluntary basis.

Scope

The International Partnership for Energy Efficiency Cooperation may include activities in the following areas:

a. Supporting the on-going work of the Participants to promote **energy efficiency** including development of nationally-determined **energy efficiency** indicators, compile best practices, and strengthen national efforts to collect data;

b. Exchanging information about measures that could significantly improve **energy efficiency** on sectoral and cross-sectoral bases such as, but not limited to:

- Standards/codes/norms and labels for buildings, **energy** using products and services with a view to accelerating the market penetration of best practices taking into account the circumstances of individual Participants;
- Methodologies for **energy** measurement, auditing and verification procedures, certification protocols and other tools to achieve optimal **energy efficiency** performance over the lifetime of buildings and industrial processes, relevant products, appliances and equipment;

- Enabling environmental tools for the financing of **energy efficiency** measures, and establishing principles for encouraging investments in energy efficiency;
- Public procurement policies for encouraging uptake of **energy** efficient products, services and technologies;
- Programs that help public institutions to become more efficient in building, vehicle, product and service purchasing and operations;
- Activities to increase the awareness of consumers and stake holders through dissemination of clear, credible and accessible information on **energy efficiency** with a view to enabling well-informed decisions;
- Best practice guidelines for evaluating the effectiveness of **energy efficiency** policies and measures;
- Public-private **cooperation** to advance **energy** efficient technology research, development, commercialization and deployment, diffusion, and transfer of such technologies;
- Actions to accelerate dissemination and transfer of best practices and efficient technologies and capacity building in developing countries;

c. Developing public-private partnerships for improving **energy efficiency** in and across key **energy** consuming sectors building on relative initiatives;
d. Enabling joint research and development into key **energy** efficient technologies;
e. Facilitating the dissemination of **energy**-related products and services that contribute to improving **energy efficiency**, and
f. Other aspects as mutually decided upon by the Participants.

The **Partnership** will provide a forum for discussion, consultation and exchange of information. It will not develop or adopt standards or **efficiency** goals for the Participants.

The present declaration does not create any legal obligation for the Participants or represent a legal obligation for the Participants or represent an **International** treaty. The Participants take part in the activities of the **Partnership** in accordance with their internal procedures.

Organizational arrangement of the Partnership

The Participants will determine the terms of their association at their first meeting by consensus, based on the principles of equitable and voluntary participation, by the end of this year.

Considering the importance of the **Partnership**, the need to ensure maximum **International** support and **cooperation**, and to be complementary with other **International** cooperative frameworks, it is envisaged that meetings of the **Partnership** will be of a high-level nature, and will meet as mutually decided but, in principle, at least once a year.

The **Partnership** may be established on the pattern of an IEA Implementing Agreement.

At least once a year, the **Partnership** will make public a summary report of its activities and plans.

IPEEC ---Declaration
6/20/08 (bold face as shown in the IPEEC declaration)

http://www.enecho.meti.go.jp/topics/g8/ipeecsta_eng.pdf

Index

A
ABESCO 292
ABN AMRO 216
Accelerated Energy Efficiency Plan 122
accountant general 131
accounting 57
    accounting regulations 320
accreditation 316
ADEME 36, 41
Adnot, Jérôme 17, 342
ADS-Maroc 102
AEPCA 311
Agency of Natural Resources and Energy (METI) 149
AGL 308
Aldridge, Alan 35
Algeria 101-106, 344
    law 102
    outlook 106
Ali, Anwar 126
alternative energy 173
alternative fuels 326
Anantasanta, Ruamlarp 147, 165, 351
ANEEL 288, 289, 290
ANME 103
APEC 259
ARCE 82
Asian Development Bank 184, 195, 207
Asia ESCO Conference 7
associations 138, 147, 261, 326
Association of Energy Engineers 138

auditing 105, 110, 121
    program 56
audits 42, 82, 103, 104, 112, 119, 120, 127, 128, 133, 185, 197, 204, 257, 274, 278, 287
Australasian Energy Performance Contracting Association (AEPCA) 307
Australia 306-311, 355
    associations 307
    barriers 310
    contracts 309
    ESCO industry development 307
    markets 308
Austria 47-49, 69
    financing 47
    government 49
    renewables 48
awareness 99

B
B.O.O.T. 179
    model 35
background 195
Balovca Hospital 70
banking sector 87
banks 82, 176, 298, 324
    commercial 76
Bannister, Paul 306, 355
barriers 22, 24, 26, 28, 36, 44, 52, 57, 60, 66, 70, 71, 84, 87, 88, 94, 98, 103, 107, 115, 121, 130, 136, 143, 167, 179, 197, 207, 225, 234, 250, 255, 282, 295, 308, 310

baseline 87, 99
 data 84
Belgium 49-53
 barriers 52
 contracts 52
 ESCO industry 51
 financing 50, 52
 future 53
 markets 49
Berliner Energie Agentur 82
Berlin Energy Agency 52
Bertoldi, Paolo viii, 17, 343, 344
best practice 308, 311
Better Buildings Partnership Loan
 Recourse Fund 233
Bishop, Rob 306
BNDES 292, 298
bonus payment 114
Boza-Kiss, Benigna 17
Brazil 286-303, 354
 barriers 295
 banks 297
 EE Program 289
 ESCO business 292
 ESCO development 287
Brazilian ESCO Association
 (ABESCO) 286
Building Greenhouse Rating 309
Bulgaria 70, 95-100, 343
 barriers 98
 contracts 97
 enabling factors 99
 ESCO industry 97
 financing 97
 history 95
Bulgarian Energy Efficiency Fund
 100
Bulgarian ESCO Fund 98
business drivers 212

Business Unit of South Africa 123

C
calculation 94
Canada 229-236, 352
 association 232
 Better Building Program 231
 contracts 232
 Federal Building Initiative 231
 financing 232
 products 233
Canadian Association of Energy
 Service Companies (CAESCO)
 232
Canadian Industry Program for
 Energy Conservation 233
carbon market brokers 254
case study 45, 67, 116, 123, 145, 202,
 246, 262, 277
Cegelec 53
CEM 28
certification 136, 316
 of ESCOs 49
Certified Energy Manager 138
chauffage 8, 13, 23, 25, 31, 49, 61,
 75, 233, 269
 contract 37
Chile 267-277, 354
 contracts 268
 financing concerns 275
 future 276
 markets 268
 obstacles 274
 policy 273
China 2, 220-227, 348
 association 221
 barriers 225
 communications 227
 opportunities 224

Index 365

types of ESCOs 223
World Bank project 221
China Energy Management
 Company Association (EMCA)
 221
Chinese ESCO association 148
CHP 23, 26, 29, 33, 35, 36, 61, 63, 81
chronic shortage 310
CIDA 194
Citi 216
Clean Tech Fund 252
climate change 116
 initiative 216
Clinton Climate Initiative 308, 311
code of ethics 132
cogeneration 16, 23, 105, 122, 308, 309
combined heat and power (CHP) 16
Comisión Nacional de Energía CNE 270
commercial banks 76
common threads 314
communications 227
Compagne Générale de Chauff (CGC) 5, 15, 27
Company Générale des Eau 27
CONADE 267
CONAE 250
conservation of natural resources 313
CONSERVE 287
constraints 207
contracts 38, 42, 52, 75, 85, 97, 109, 114, 120, 135, 161, 198, 199, 205, 222, 232, 268, 282, 309, 310
 models 31
contracting 120
contract energy management

(CEM) 27
cost of energy 67
Côte d'Ivoire 106-111, 345
 barriers 107
 contracts 109
 ESCO Project 107
 financing 108
 markets 197
 obstacles 110
 projects 110
counter measures for global warming 154
credibility 24
Croatia 88-89
 barriers 88
CS2E 41
current energy service market 216
customer awareness 324
customer perception 325
Czech Republic 59-60, 69-73
 barriers 71
 first project in Eastern Europe 70

D

Dagleish, Braam 118
Dalgleish, Braam 345
Dalkia 28, 31, 50, 53, 74, 308
Dalkia, undisclosed third-party financing 321
Dang Son, Ha 148, 352
Danish Offshore Industry 59
DEDE 168, 169
DEDE/EGAT 169
demand-side intervention 61
demand-side management (DSM) 12
demonstration projects 150, 189
Denmark 16, 59-60

barriers 60
financing 59
private sector 59
deterrents 77
Deutsche Bank 216
Development Credit Authority 141
Dexia 50, 53
DFID 194
Directive 2006/32/EC 79
Display Energy Certificates (DECs) program 30
dispute avoidance 150
district heating 36, 48
dominated by state government 309
Dow Chemical 69
Dreessen, Thomas K. 259, 321
DSCL 148
DSCL Energy Services Company 196
DSM 118, 119, 120

E
early motivation 166
EBRD 81, 90, 97, 98, 100
EC 7, 96
eco-renovation 51
Econoler viii, 148, 231, 268
Econoler International 102, 112, 126, 130, 194, 196, 229
economic growth 313
ECO Opole 75
EE 3
financing 191
Efficiency Valuation Organization (EVO) 318
EGAT 168, 169
Egypt 125, 126, 137-146, 346
association 137

barriers 143
ESCO industry profile 139
financing 140
future 144
history 137
markets 142
potential 139
projects 142
Egyptian Energy Service Business Association (EESBA) 137
electricity 300
prices 270
El Khoury, Pierre 126
EMCA 148, 220
emissions trading 311
EMSTAR 27
enabling factors 99, 121, 259, 285
energy 2, 67
audits 62, 63, 65, 213
Energy Business Reports 16
Energy Center in Bratislava 86
energy certification 87
regulation 62
Energy Conservation Act 196
Energy Conservation Systems 308
energy costs 310
energy efficiency 3, 53, 88, 109, 122
Energy Efficiency Act 99
energy efficiency action plan (EEAP) 79
Energy Efficiency Building Retrofit Program (EEBRP) 216
Energy Efficiency Fund 106, 281
energy efficiency investments 57, 104
energy efficiency plan 62
energy intensity 96
Energy Line 50
energy management 222

Index

energy performance contracting (EPC) vii, 1, 6, 7, 15
Energy Regulatory Commission 116
Energy Saving Partnership (ESP) 20
Energy Sector Management Assistance Program (ESMAP) 260
energy services agreements 257
energy service companies (ESCO) vii, 1, 4, 7
energy service providers (ESPs) 7
Energy Solutions 306
energy suppliers 61
energy supply 65, 97
Enfinity 51
environmental benefits 301
Envo 134
Envo/GESI 135
Envo Energy Services 132
EPC 6, 8
EPS 70
EPS Capital 260
equipment 40, 44, 53, 57, 72, 81, 92, 105
equipment suppliers 60
ESCO 1, 4
ESCOPhil 204
ESCOs vii
    association 325
    Europe 7
    industry 30, 41, 107, 138, 204, 250, 279
    market 107
    ownership 239
    pilot project 170
    projects 29, 160
    services 8
    support 187
    types of 223
ESCO Update Report of 2007 17
ESENER S.A. 268
Eskom Energy Crisis Committee 118
Estonia 69
EURCONTRACT 52
European Bank of Reconstruction and Development 74, 89
European Commission 7, 343, 344
European Commission's Green Light Programme 25
European ESCO Status Report 2005 16
European market 15
European Union 18, 89
EVO 259, 320
Excellent Energy International Company, Ltd. 171
Export Import Bank 252

F
facilitators 62
Facility for Municipal Energy Efficiency 98
facility management contracts 37
FBI 233
Federal Building Initiative (FBI) 231
federal law 89
federal level 91
federal program 91
Fedesco 49
Ferreira, Marcelo Ulises Gonzalez 355
FG3E 41, 44
financial 272
    institutions 60, 76, 97

instruments 78
mechanisms 66
models 9
support 121
financing 31, 38, 47, 50, 52, 58, 59,
    61, 66, 75, 84, 86, 93, 97, 107,
    108, 114, 120, 135, 140, 199, 219,
    232, 242, 251, 281, 284, 319
    concerns 275
    energy efficiency 26, 56
    lack of 103
    mechanisms 321
Finland 55-57
    barriers 57
    financing 56
    industries 55
    opportunities 57
    public sector 56
first out 88, 234, 282, 322
FIs 25
Fondelec Latin-American Clean
    Energy Services Fund 252
forecast 95
France 36-47, 341
    barriers 44
    case study 45
    contracts 56
    ESCO industry 41
    financing 38
    history 36
    markets 41
    projects 43
    services provided 42
French Dalkia 75
Fund for Environmental Protection
    and Energy Efficiency 88
Fund for the Reduction of the
    Global Energy Cost 51
future 137, 144, 182, 200, 276, 300

G
Gas Distribution 233
Gauer, Karin 267, 354
GDP 211, 225
GEF 75, 79, 81, 88, 103, 104, 106,
    127, 168, 196, 220, 221, 225, 279,
    281, 297, 302
GEF-KAM 112, 113
GEF/UNDP 112
Germany 15, 18-22
    barriers 22
    growth factors 20
    government 21
German Carbon-Aid Fund 83
German MVV 75
GHG 215, 216, 225
Gilligan, Donald 229, 237, 354
global energy market 2
Global Environmental Facility 118,
    126, 220
Global Opportunities Fund 259
global warming 311
Gonzalez, Marcelo 267, 278
Good, Larry 126, 132, 347
government 79, 129, 131, 134, 173,
    197, 234, 308, 310, 314
    action 86
    regulations 36
    energy policy 116
    procurement 155
    resolution 90
    role 314
    support 108
Government of India 194
Graz Energy Agency 47
Greece 25-26, 70
    barriers 26
    financing 26
    government 26

Index 369

greenhouse gas (GHG) 3
green certificates 50
Green Invest 53
green procurement 273
Green Procurement Law 153
GREI 278
Grobler, L.J. 101, 118, 346
Gross, Z'ev 126, 346
growth 158
GTZ 194
guaranteed savings 6, 9, 10, 31, 52, 65, 71, 109, 114, 120, 130, 135, 160, 162, 197, 202, 206, 219, 232, 251, 321
guaranteeing savings 65
guarantee fund 109
Güven, Halil 126, 132, 347

H
Hanneman Hospital 8
Hansen, Jim viii
Hansen Associates viii, 70, 118, 148, 195, 196, 221, 305
Hassan, Emad 126, 346
Herrera, Alice B. 148, 351
history 90, 112, 238
Honeywell 50, 53, 216, 308
housing sector 93
human resources, lack of 310
Hungary 83-86
    barriers 84
    opportunities 85
Hydro-Quebec 102
hydropower 106

I
ICICI Bank 200
IEEFP 230, 261
IEEFP-Mexico 253, 259, 262

IEPF 107, 110
IGAs 130
Ikaros Solar 51
improved energy efficiency 313
incentives 38, 39, 41, 49, 52, 61, 63, 104, 105, 130, 136, 233, 307
India 2, 194-203, 349
    association 194
    background 195
    banks 200
    barriers 197
    case study 201
    financing 199
    future 200
    measurement and verification 198
    private sector 196
Indian Council for Promotion of Energy Efficiency Business 194
Indian Renewable Energy Development Agency 194, 318
industrial energy efficiency 183
industrial plants 93
industrial sector 63, 84, 85
industry 55, 82, 97
    drivers 311
    sectors 91
information 72, 73, 84, 85, 86, 103, 109, 274, 277, 282, 284
    dissemination 151
infrastructure 91
innovative technologies 173
integrated 310
    solution 13
Inter-American Development Bank 268
interest rate 275
international aid 84
International Energy Agency (IEA)

2, 133
International Energy Efficiency Finance Protocol (IEEFP) 259, 320
International Financial Corporation 71
International Partnership for Energy Efficiency Cooperation 4
International Performance, Measurement and Verification Protocol (IPMVP) 119, 238, 244, 257, 318
introductory study 149
investment recovery 149
IPEEC 4
IPMVP 122, 238, 318
Iqbal, Anees 17, 27, 342
IREDA 148, 194, 196
Ireland 16, 35-36
    barriers 36
Israel 126, 129-132, 346
    government 131
    guaranteed savings 130
    pilot project 131
    shared savings 132
Israeli law 129
Israeli Ministry of National Infrastructure (MNI) 126
Italy 23-25
    barriers 24
    government 23

J
Japan 148-165, 349
    association 151
    contracts 150, 161
    global warming 154
    growth 158

    introductory study 149
    shared savings 160, 165
Japanese Association of Energy Service Companies (JAESCO) 147, 151
Japan External Trade Organisation (JETRO) 184
Japan International Cooperation Agency (JICA) 133
Johnson Controls 50, 53, 216, 308
Jordan 125
JP Morgan Chase 216

K
KAM 113
KEMCO 212
Kenya 112-117, 345
    barriers 115
    case studies 116
    contracts 114
    enabling factors 115
    ESCO development 112
    financing 114
    history 112
    markets 113
    outlook 116
Kirai, Paul 101, 112, 345
Korea 350
    Energy Management Corporation 212
Korean Association of ESCO Companies 218
Kumar, D.V. Satya 349
Kyoto 55
Kyoto Protocol 133, 151, 154, 225, 254, 272, 307

L
lack of financing 103

Index

Landys & Gyr 74
Langlois, Pierre viii, 229, 327, 352
laws 85, 95, 225
LCEC 127, 128
LCECP 126
leasing 97
Lebanon 125-129
    LCEC 127
    training 128
legal 62, 66, 70, 77
    ambiguities 57
    and institutional environment 26
    and regulatory frameworks 16
    environment 171
    framework 61
    guidelines 193
    uncertainties 22
legislation 50, 85, 88, 96
    changes 26
Light's Labours Lost 4
light emitting diode (LED) 122
Light of Our Eyes 83
Lim, Jim K.Y. 182
Lithuania 70
LNG 125
load shifting 122
loan guarantee program 221
local financial institutions 254
Longhai, Shen 148, 220, 349
loose contracts 171
low interest rate loan 175, 176
Luxembourg 55

M
M&V 84, 99, 138, 143, 145, 168, 290
    limitations 245
maintenance 39, 40, 45, 64, 65
    contracts 52

Malaysia 182-192
    ESCO support 187
    financing 191
    Industrial EE Improvement Program (MIEEIP) 183
Malaysian Energy Professionals Association (MEPA) 187
Manchev, Pavel 95, 343
manufacturers programme 190
market 41, 75, 79, 92, 113, 135, 136, 142, 155, 172, 212, 232, 239, 250, 258, 268, 279, 308, 309
    constraints 245
    development 270
    drivers 241
    segments 31, 44, 120
        types of 105
    survey 155
Masuda, Takashi 147, 350
measurement and verification (M&V) 17, 21, 103, 151, 139, 193, 198, 244, 257, 260, 276, 297, 311, 314, 317
Memik, Ata Osman 126, 132, 348
memorandum of understanding 42
Menon, Nisha 148, 196, 349
MEW 127
Mexico 249-266, 353
    association 261
    barriers 255
    case studies 262
    enabling factors 259
    ESCO industry 250
    financing 251
    markets 250, 258
    projects 255
    regulatory constraints 258
micro-wind 63
MNI 130, 131, 132

model contract 308
MoEN 175, 176
MOIE 103
monitoring and verification 73, 94, 110
Morocco 101, 106, 344
    enabling factors 105
    outlook 106
Motiva 55
municipalities 61, 93, 121
municipality 96
municipal authorities 84
municipal level 91
municipal sector 87
Murakoshi, Chiharu 147, 349
MUSH 240
MVV 74

N
N'Guessan, R. M'Gbra 101, 345
NABERS Energy 309
Nacional Financiera (NAFIN) 252
NAFIN 253, 260, 261
Nakagami, Hidetoshi 147, 350
National Accounting System 193
National Association of Energy Service Companies (NAESCO) 237
National Bank of Public Works and Services 253
National Britannia 134
National Electricity Response Plan 122
National Energy Commission 270
National Energy Consultants 127, 128
National Energy Efficiency Agency (NEEA) 119
national energy service company 95
NEDO 150
negotiation 106
Netherlands 16, 54
    competition 54
    markets 54
    projects 54
new beginning 174
New South Wales 307
New Zealand 305-306
    government involvement 305
    tariff changes 306
Nexant 126
no-interest loan 108
non-OECD 2
northern Africa 101
North American Development Bank 261
North American Development Bank (NADBank) 253
no trust 67
NREL 250
NREL-CONAE 262

O
O&M personnel 30, 168
obstacles 274, 313
open book-type contracts 32
operating and manning 27
operating contracts 38
operation and maintenance 30, 36, 37, 43
opportunities 85, 87, 179, 224
Organization of Economic and Cooperative Development (OECD) 3
outlook 116, 311
outsourcing 64
ownership 78

Index

P
P1 contracts 38
P2 contracts 39
P3 type contracts 40
Palestinian Authority 125
partial credit guarantees 98
Partial Loan Guarantee
    Programme 79
partial risk guarantee fund 104
payments 94
PC projects 160
PECU 19, 22
Peoples Republic of China 148
perceived uncertainty 61
perceived value 173
Perez, Monica 230, 249, 320, 353
PFI 29, 30, 31
Philippines 204-210, 351
    association 204
    barriers 207
    contracts 205
    ESCO industry 204
    way forward 208
pilot 143, 145
    projects 131
POE ESCO 74, 75
Poland 70, 73-80
    case studies 80
    contracts 75
    ESCO development 73
    financing 75
    limitations 77
    markets 75
    projects 76
policy 225, 273
pollution 214
Poole, Alan Douglas 267, 286, 354
Pop, Florin 82
Portugal 16, 60-63

    barriers 61
    facilitators 61
    government 61
potential 139
power conservation programme 122
power factor 145
PPEE 270, 271
preventive maintenance 65
Private-Public Partnership (PPP)
    258, 277
Private Financing Initiative (PFI) 28
private sector 59, 72, 93, 196
private service sector 48
PROCEL 287
procurement 57, 310
    regulations 143
    rules 285
products 233
PROESCO 292, 293, 302
projects 135, 142, 221, 255, 282
    economically unattractive 66
project financing 104
project types 76, 94
Prokopiev, Dichko 97
public-private partnership (PPP)
    15, 43, 86, 277
public agencies 63
public buildings 274
Public Internal Performance
    Commitments (PICO) 21
public lighting 88
public sector 56, 58, 62, 72, 92, 142,
    194, 197, 294
Public Sector Energy Efficiency
    Project 83, 85

R
Rational Energy Utilization Act 210
REEEP (Renewable Energy

Efficiency Partnership) 272
registration 132
regulations 78, 99, 134, 225, 277
regulatory constraints 23, 258
RELUZ 290
renewables 61, 102, 240, 242, 259, 261, 306, 307, 326
   energy 25, 45, 48, 63, 84, 133, 173
Republic of Korea (South Korea) 210-219
   association 218
   audits 213
   business drivers 212
   Climate Change Initiative 216
   contracts 213
   financing 219
   government 212
   markets 216
   pollution 214
RES 26, 58
residential sector 294
return rates 61
revolving fund 108
Rezessy, Silvia 17
RFPs 145, 197
risk-sharing credit window 293
risk 150, 176, 179
   management 171
   mitigation 132
Rokhlikov, Igor 89, 344
Romania 70, 81-83
   banks 82
   EE Law 81
   industry 82
   markets 82
   UNDP/GEF 81
Romanian Energy Efficiency Law 81
Roncero, Enrique Gonzalez 63

Rosenstein, Frédéric 17, 36, 342
Royal Dutch Shell 5
Russia 88-95, 344
   barriers 94
   financing 93
   forecast 95
   history 90
RWE Solutions 96

S
Satya Kumar, D.K. 148
savings guarantee 222
Scallop Thermal 5
Schneider Electric 40
seed capital 108
Senegal 106
SEVEn 70, 71
Shahin, Walid 125
shared 135
shared savings 5, 6, 9, 10, 12, 31, 47, 61, 65, 97, 110, 114, 132, 160, 162, 165, 177, 199, 200, 202, 206, 219, 222, 233, 249, 251, 279, 282, 306, 321
Shell UK 27
Siemens 50, 53, 70, 74, 216
Siemens, Johnson Control 40
Siemens Building Technology 210
Singer, Terry 229, 237
skilled personnel 245
Slovakia 69, 86-87
   barriers 87
   opportunities 89
SMEs 105, 121
SME Guarantee program 253
solar 48, 63
   thermal 66
South Africa 118-123, 345
   association 119

barriers 121
case studies 123
contracts 120
enabling factors 121
financing 120
future 122
markets 120
South African Association of ESCOs (SAAE) 119
South Korea 210
Spain 63-68
    barriers 66
    case study 67
    contracts 65
    ESCO industry 65
    future 67
    incentives 63
    markets 65
    projects 65
    public agencies 63
special purpose company 179
special purpose entities 12, 251, 260, 321
special purpose vehicle 12
split incentives 44, 48, 61, 85
SPV 12
staff development 137
standard contracts 150
standard procedures 62
street lighting 96
Sub-Saharan Africa 106
subsidized 87
subsidy 78, 83
Sugay, Ed 148, 210, 350
supply-side investments 61
sustainability vii, 246
sustainable development 88, 224
Sustainable Energy Ireland (SEI) 35
sustainable energy use 21

Sweden 57-59
    market 58
    opportunities 59
    programs 58

T

tariffs 105, 116, 306
tax law 78, 320
TBE Chile 268
technical 272
technologies 163
tender 130
Thailand 165-182, 351
    banks 176
    barriers 167, 179
    early motivation 166
    legal environment 171
    pilot project 168 – 172
    shared savings 177
    special purpose company 179
The Netherlands 54
third-party financing 50, 52, 99
third-party investors 53
Three Country Energy Efficiency Project 197
Total Energy Systems 308
TPF 23, 25, 26, 37, 50, 59
TPF-Econoler 49
trade association 30
trade receivables fund 299
training 128, 133, 137
Trane 216
transversal measures 53
trigeneration 308, 309
Triodos 53
trust 22, 99
Trust Fund for Electric Energy Savings (FIDE) 252
Tunisia 101-106, 344

barriers 103
market 105
outlook 106
prospect 105
Turkey 126, 132-137, 347
   barriers 136
   contracts 135
   enabling factors 136
   financing 135
   government 134
   IEA 133
   market 135
   projects 135
Turkish Directorate of Electrical Power Resources Survey and Development Administration (EIE) 133
Turkish Electricity Authority 133
turnkey 110

U
UBS 216
UEEP 279
undisclosed third-party financing 31
UNDP 81, 88, 106, 107, 113, 125, 127, 137, 139, 145, 302
UNDP's credit guarantee 141
UNDP-GEF-DOE PELMAT 207
UNDP/GEF 85
UNDP/GEF Energy Efficiency Project 81
UNECE 96
UNEP 125
Union Fenosa 63
United Kingdom 27-35, 342
   association 30
   barriers 28
   case studies 33

   factors 32
   financing 31
   history 30
   market 31
   projects 29
United Nations Development Program 126
United States 237-249, 354
   case studies 246
   ESCO ownership 239
   financing 242
   history 238
   market 239, 241
   market constraints 245
   measurement and verification 244
universities 120, 307
Uruguay 278-286, 355
   barriers 282
   contracts 282
   enabling factors 285
   ESCO industry 279
   financing 281
   market 279
   projects 282
Uruguayan Energy Efficiency Fund 286
Uruguay Energy Efficiency Project 282
USAID 81, 98, 102, 126, 134, 137, 138, 140, 141, 145, 195
US AID 194
US Trade Development Agency 260
UTE 278
UTE-USCO 280
utilities 24, 37, 50, 52, 61, 115, 122, 168, 233, 238, 239, 242, 244, 252, 278, 287, 288, 289, 290, 302, 303, 305

Index

utility/grid companies 60

V
value chain 12
Vanstraelen, Lieven 17, 341
Vechakij, Arthit 147, 165, 351
Veolia Environnement 268
verified savings 115
Vietnam 192-193, 352
    company categories 192
    contracts 193
Vivendi 28, 268

W
way forward 208

western Europe 15
white certificates 24, 45
widely recognized 313
World Bank 73, 75, 76, 79, 81, 88, 89, 103, 107, 148, 193, 194, 196, 197, 220, 225, 260, 279, 290, 318
World Bank GEF 169

Y
Y Lim, Jim K. 350

Z
Zahar, Hakim 101, 344
Zhao, Ming (Lily) 148, 220, 348